THE EVOLUTION OF ARTIFICIAL INTELLIGENCE

FROM VIRTUAL ASSISTANTS TO AUTONOMOUS ROBOTS

DAVID SANDUA

The evolution of Artificial Intelligence.

"Artificial intelligence will change the world, but the question is how.

Charlie Rose

INDEX

7

8

I. INTRODUCTION

Artificial intelligence, a concept that once seemed like science fiction, has now become an integral part of our daily lives. From virtual assistants like Siri and Alexa to autonomous robots performing complex tasks, AI has rapidly evolved and permeated various sectors. This evolution has revolutionized how we interact with technology and has pushed the boundaries of what machines can achieve. The transition from virtual assistants to autonomous robots represents a significant milestone in the development of AI, marking a shift towards more sophisticated and independent systems that can operate in real-world environments. Virtual assistants were among the first manifestations of AI that gained widespread popularity, offering users the ability to interact with technology using natural language commands. These virtual assistants, equipped with speech recognition and natural language processing capabilities, have become indispensable tools for tasks such as setting reminders, searching for information, and controlling smart home devices. The impact of virtual assistants extends beyond personal use, with industries leveraging their capabilities to improve customer service, streamline operations, and enhance productivity. As technology continued to advance, the focus shifted towards developing autonomous robots that could operate autonomously without human intervention. These robots, ranging from autonomous vehicles to industrial machines, are equipped with sensors, actuators, and AI algorithms that enable them to navigate complex environments and perform tasks efficiently. While the development of autonomous robots presents exciting opportunities for innovation, it also raises ethical and societal concerns

10

regarding issues such as job displacement, safety, and privacy. As AI continues to evolve, it is crucial to strike a balance between technological progress and ethical considerations to ensure that these advancements benefit society as a whole.

Definition of Artificial Intelligence

Virtual assistants have been instrumental in shaping the modern landscape of artificial intelligence. These intelligent software programs are designed to assist users in various tasks, from setting reminders to answering complex queries. Companies like Apple, Amazon, and Google have spearheaded the development of virtual assistants such as Siri, Alexa, and Google Assistant, which have become integrated into everyday life for millions of people. The initial functionalities of virtual assistants were centered around speech recognition and basic commands, but advancements in natural language processing have allowed these systems to understand and respond to more complex interactions with users. Through the integration of machine learning and neural network technologies, virtual assistants have evolved to provide personalized responses and recommendations, making them invaluable tools in both personal and professional settings. As virtual assistants continue to improve in their capabilities, the focus has shifted towards developing autonomous robots - the next frontier of artificial intelligence. Autonomous robots are defined by their ability to operate independently, making decisions and performing tasks without human intervention. Examples of autonomous robots include self-driving cars, cleaning robots, and industrial robots used in manufacturing processes. These machines are equipped with advanced sensors, artificial intelligence algorithms, and sophisticated control systems that enable them to navigate complex environments and adapt to changing conditions. While the development of autonomous robots holds great promise for revolutionizing industries such as transportation, healthcare, and agriculture, there are also significant technological and ethical challenges that must be addressed, including safety concerns, data privacy, and the impact on the workforce. The proliferation of autonomous robots in various sectors underscores the growing impact of artificial intelligence on society. From assisting in medical procedures to optimizing supply chains in logistics, these intelligent machines are reshaping industries and transforming the way we live and work. As businesses increasingly adopt autonomous robots to streamline operations and increase efficiency, it is essential to consider the potential benefits and risks associated with their widespread use. While the deployment of autonomous robots has the potential to enhance productivity and drive innovation, there are also concerns about job displacement, ethical dilemmas, and the need for regulatory frameworks to govern their deployment. By exploring the evolution of artificial intelligence from virtual assistants to autonomous robots, we gain insights into the technological advancements that are shaping our future and the critical considerations that must be addressed to ensure that AI benefits society as a whole.

Brief History of Artificial Intelligence

Technological Progress: From Speech Recognition to Natural Language Understanding. As artificial intelligence continued to advance, significant progress was made in the field of natural language processing, allowing virtual assistants to move beyond simple speech recognition towards a more sophisticated level of understanding. Through the integration of machine learning and neural network technologies, virtual assistants such as Siri, Alexa, and Google Assistant were able to not only recognize spoken words but also interpret and respond to complex commands in a more human-like manner. This development marked a crucial step in the evolution of AI, as it enabled virtual assistants to interact with users in a more intuitive and efficient way, revolutionizing the way people access information, perform tasks, and communicate with technology. The next phase in the evolution of artificial intelligence saw the emergence of autonomous robots, which represented a significant leap forward in the capabilities of AI-powered machines. Autonomous robots are characterized by their ability to perform tasks and make decisions independently, without direct human intervention. Examples of current autonomous robots include autonomous vehicles, cleaning robots, and robots used in industrial settings. These robots are equipped with sensors, cameras, and advanced algorithms that enable them to navigate their environment, interact with objects, and carry out specific functions with a high degree of autonomy. The development of autonomous robots also poses technological and ethical challenges, as the integration of AI in robots raises questions about safety, privacy, and ethical considerations that must be addressed as these technologies become more prevalent in society.

Thesis Statement

The evolution of artificial intelligence has taken significant strides from the early days of virtual assistants to the more complex autonomous robots of today. Virtual assistants, such as Siri, Alexa, and Google Assistant, marked the initial foray into AI, providing users with basic functionalities like setting reminders, playing music, and answering simple questions. These early developments laid the foundation for more sophisticated natural language understanding capabilities, enabling virtual assistants to process complex commands and engage in more meaningful interactions with users. The impact of virtual assistants extends beyond personal convenience, influencing various industries by enhancing customer service, streamlining operations, and driving innovation in communication technologies. As technology progressed, the focus shifted towards developing autonomous robots capable of performing tasks independently without human intervention. Autonomous robots, like autonomous vehicles and cleaning robots, exhibit a higher degree of intelligence and automation, making them valuable assets in industries ranging from manufacturing to healthcare. The advancement of autonomous robots also presents challenges, both technological and ethical, such as ensuring safety in autonomous vehicles and addressing concerns about job displacement due to automation. Despite these challenges, the potential benefits of autonomous robots in terms of efficiency, productivity, and safety are driving further research and development in the field of robotics. Looking ahead, the future of artificial intelligence holds promising trends and perspectives that are poised to shape society in profound ways. From the integration of AI with other cutting-edge technologies like blockchain and quantum computing to the rise of ethical considerations in AI development, the trajectory of AI evolution is multifaceted and dynamic. As we navigate this rapidly changing landscape, it is crucial to strike a balance between technological innovation and ethical considerations to ensure that AI continues to benefit humanity without compromising fundamental values and principles. The journey from virtual assistants to autonomous robots underscores the transformative power of AI and highlights the need for thoughtful reflection and proactive governance to steer its course responsibly.

II. VIRTUAL ASSISTANTS

As virtual assistants have become increasingly integrated into our daily lives, they represent a significant milestone in the development of artificial intelligence. These AI-powered programs, such as Siri, Alexa, and Google Assistant, have revolutionized the way we interact with technology by providing personalized assistance through voice commands. Initially designed for simple tasks like setting alarms or sending messages, virtual assistants have evolved to offer a wide range of capabilities, including answering complex questions, providing recommendations, and even controlling smart home devices. Their impact extends beyond individual users, as businesses have also embraced virtual assistants to streamline customer service and improve operational efficiency. The technological progress that has enabled virtual assistants to advance from basic speech recognition to sophisticated natural language understanding has been a key driver in their evolution. Breakthroughs in natural language processing have empowered virtual assistants to not only recognize words but also comprehend context, tone, and intent behind user queries. By leveraging machine learning algorithms and neural networks, these AI systems continuously learn and improve their responses, making interactions more seamless and intuitive. This transformation has significantly enhanced the usability and effectiveness of virtual assistants in a variety of tasks, from providing personalized recommendations to facilitating hands-free communication. Looking ahead, autonomous robots represent the next frontier in artificial intelligence, combining the capabilities of virtual assistants with physical mobility and autonomy. These robots, such as autonomous vehicles, cleaning robots, and industrial automation systems, have the potential to revolutionize various sectors by performing complex tasks independently and efficiently. The development of autonomous robots also poses significant technological and ethical challenges, including ensuring safety, privacy, and accountability in their operations. As we navigate towards a future where autonomous robots play a more prominent role, it is crucial to address these challenges in order to harness the full potential of AI while upholding ethical standards and societal values.

Definition and Functionality

Technological advancements have led to the development of autonomous robots, which represent the next frontier of artificial intelligence. These robots are defined as machines that can perform tasks or work independently without human intervention. They often incorporate a combination of sensors, actuators, and AI algorithms to navigate their environment and make decisions. Examples of autonomous robots include self-driving cars, cleaning robots, and industrial robots used in manufacturing processes. The emergence of autonomous robots raises both technological and ethical challenges that need to be carefully addressed to ensure their safe integration into society. The functionality of autonomous robots extends beyond mere automation as they possess the ability to perceive and respond to their surroundings in real-time. These robots are equipped with advanced sensors such as cameras, Lidar, and Radar to collect data about their environment. This data is then processed by AI algorithms to make decisions autonomously. Autonomous robots can adapt to changing situations, learn from experience, and optimize their performance over time. Their potential applications span a wide range of industries, from healthcare to agriculture, offering new possibilities for improving efficiency and productivity. While the use of autonomous robots offers numerous benefits, their widespread adoption poses significant risks that need to be carefully considered. These risks include concerns about data privacy, cybersecurity vulnerabilities, and the potential displacement of human workers in certain industries. Ethical dilemmas surrounding the use of autonomous robots, such as issues related to accountability and decision-making, also need to be addressed. It is essential for policymakers, researchers, and industry stakeholders to collaborate in order to develop robust regulations and guidelines that ensure the responsible deployment of autonomous robots in society.

Popular Virtual Assistant Technologies

Virtual assistants have become an integral part of modern life, with technologies such as Siri, Alexa, and Google Assistant leading the way in AI interactions. These virtual assistants have evolved from basic voice-command devices to sophisticated AI systems capable of understanding and responding to complex queries. Their impact on everyday life is profound, simplifying tasks and providing convenience in various industries such as healthcare, finance, and entertainment. The development of virtual assistants has paved the way for advancements in natural language understanding, enabling these systems to interpret and respond to human language with increasing accuracy and efficiency. Through the integration of machine learning and neural network technologies, virtual assistants have made significant strides in their capabilities, moving beyond simple command execution to context-aware interactions. The progression from speech recognition to natural language understanding has been key in enhancing user experiences and expanding the applications of virtual assistants. This evolution has also raised ethical considerations regarding privacy and data security, as these systems collect and analyze vast amounts of user information to improve their performance. As AI continues to develop, ensuring the ethical use of these technologies will be crucial in maintaining trust and promoting responsible innovation. Looking ahead, the next phase in AI evolution is the emergence of autonomous robots, which combine AI technologies with physical mobility to perform tasks independently. From autonomous vehicles to robots in industries such as manufacturing and agriculture, these technologies offer new possibilities for automation and efficiency. Challenges related to safety, regulation, and societal impact need to be addressed as autonomous robots become more prevalent. As AI continues to advance, the synergy between virtual assistants and autonomous robots will shape the future landscape of technology, transforming the way we work, interact, and live in a rapidly evolving digital world.

Impact on Daily Life

The impact of artificial intelligence on daily life is undeniable, with the evolution from virtual assistants to autonomous robots reshaping how we interact with technology. Virtual assistants like Siri, Alexa, and Google Assistant have become integral parts of our routines, providing convenience and efficiency in accessing information, managing tasks, and controlling smart devices.

The ability to converse with these AI-driven systems has fundamentally altered the way we navigate our daily lives, blurring the lines between human and machine interactions. As virtual assistants continue to evolve with advancements in natural language processing and machine learning, their presence in our lives will only grow more pervasive, influencing everything from how we shop and communicate to how we work and entertain ourselves. Moving beyond virtual assistants, the emergence of autonomous robots represents a new frontier in AI with profound implications for daily life. These robots, equipped with the ability to operate independently and make decisions based on their surroundings, have the potential to revolutionize various industries, from transportation and healthcare to manufacturing and agriculture. The integration of autonomous robots into our everyday lives offers exciting possibilities for increased efficiency, productivity, and safety. The development of such advanced systems also brings forth significant technological and ethical challenges, raising important questions about the impact on employment, privacy, and societal well-being. As autonomous robots continue to be implemented in real-world applications, the impact on daily life becomes more tangible, with examples ranging from self-driving cars and delivery drones to robotic surgery and automated manufacturing processes. The use cases for autonomous robots span across diverse sectors, promising increased reliability, precision, and cost-effectiveness in various operations. While these technological advancements have the potential to improve our quality of life and drive economic growth, they also raise concerns about job displacement, data security, and the ethical ramifications of handing over decision-making power to machines. It is crucial to carefully consider the benefits and risks associated with the widespread adoption of autonomous robots, ensuring that these innovative technologies are deployed responsibly to maximize their positive impact on society.

III. MACHINE LEARNING

The integration of machine learning techniques has played a pivotal role in advancing the capabilities of artificial intelligence, enabling machines to learn from data and make decisions without explicit programming. Machine learning algorithms have been instrumental in enhancing the accuracy, efficiency, and adaptability of AI systems, allowing them to evolve and improve over time. By utilizing neural networks and other sophisticated mathematical models, machine learning has transformed the way AI systems are developed and deployed, enabling them to perform complex tasks such as natural language processing, image recognition, and autonomous decision-making. One of the key benefits of machine learning in the context of artificial intelligence is its ability to continuously learn and adapt to new information. This adaptive learning process allows AI systems to analyze vast amounts of data, identify patterns, and make predictions based on new inputs. Through techniques such as reinforcement learning and deep learning, machine learning algorithms can optimize their performance and make more accurate decisions over time. This iterative learning process is essential for autonomous robots, as it enables them to navigate dynamic environments, interact with humans, and perform complex tasks with minimal human intervention. As machine learning continues to advance, the possibilities for artificial intelligence and autonomous robotics are limitless. From self-driving cars to intelligent industrial robots, the integration of machine learning algorithms has the potential to revolutionize various industries and pave the way for a future where intelligent systems can work alongside humans seamlessly. While there are challenges and ethical considerations to address, the evolution of machine learning and artificial intelligence holds great promise for enhancing efficiency, productivity, and innovation in the years to come.

Explanation of Machine Learning

As machine learning has become increasingly prevalent in various fields, it is essential to understand its fundamental principles. Machine learning refers to the subset of artificial intelligence that enables computers to learn and improve from experience without being explicitly programmed. Through the use of algorithms and statistical models, machines can analyze data, identify patterns, and make decisions. One of the key aspects of machine learning is its ability to adapt and improve over time as it processes more information. This iterative process allows machines to refine their predictions and optimize their performance based on feedback. Machine learning techniques can be categorized into several types, including supervised learning, unsupervised learning, and reinforcement learning. In supervised learning, machines are trained on labeled data to make predictions or classifications. Unsupervised learning involves extracting patterns and insights from unlabeled data using techniques such as clustering and dimensionality reduction. Reinforcement learning, on the other hand, focuses on training agents to make sequential decisions by rewarding or punishing their actions based on their outcomes. Each of these approaches has its unique advantages and applications in different domains. The rapid advancements in machine learning have led to the development of sophisticated algorithms and models that can tackle complex problems across diverse industries. From image recognition and natural language processing to autonomous navigation and predictive analytics, machine learning has revolutionized how we interact with technology and process information. As researchers continue to push the boundaries of what is possible with AI, it is crucial to consider the ethical implications and potential biases that may arise from the use of machine learning algorithms. By understanding the underlying principles and limitations of machine learning, we can harness its power to drive innovation and address some of the most pressing challenges facing society today.

Types of Machine Learning Algorithms

Advances in machine learning have paved the way for the development of various types of algorithms that underpin artificial intelligence systems. Supervised learning algorithms, for instance, are designed to learn from labeled data, allowing machines to make predictions or decisions based on this training. This type of algorithm is commonly used in applications like image recognition, spam detection, and recommendation systems. Unsupervised learning algorithms, on the other hand, do not rely on labeled data but instead identify patterns and relationships within unlabeled datasets. Clustering algorithms, such as K-means and hierarchical clustering, belong to this category and are often employed in tasks like customer segmentation and anomaly detection. Reinforcement learning algorithms represent a distinct approach in machine learning, wherein an agent learns to interact with an environment through trial and error. By receiving rewards or penalties based on its actions, the agent continuously refines its decision-making process to achieve a specific goal. This type of algorithm has been pivotal in the development of autonomous systems, such as self-driving cars and robotic arms. Deep learning algorithms, inspired by the structure and function of the human brain, have gained prominence in recent years for their ability to process large volumes of data and extract complex patterns. Convolutional neural networks (CNNs) and recurrent neural networks (RNNs) are examples of deep learning architectures widely used in tasks like image and speech recognition. While each type of machine learning algorithm has its strengths and weaknesses, the overall progress in AI has been accelerated by the combination and refinement of these techniques. Hybrid approaches, such as ensemble learning, where multiple models are combined to improve accuracy, have become increasingly popular in complex AI applications. As researchers continue to push the boundaries of machine learning and AI, the future holds promising advancements that will further redefine the capabilities and impact of intelligent systems in various domains.

Applications in Artificial Intelligence

Advancements in artificial intelligence have paved the way for a wide array of applications across various industries, revolutionizing the way tasks are performed and decisions are made. One key application that has significantly impacted daily life is virtual assistants, such as Siri, Alexa, and Google Assistant. These virtual assistants have evolved from basic voice command recognition systems to sophisticated tools that can understand and respond to natural language queries. Their integration with machine learning algorithms and neural network technologies has enhanced their accuracy and effectiveness, making them indispensable in tasks ranging from setting reminders to conducting online searches. The widespread adoption of virtual assistants in homes, businesses, and healthcare settings has reshaped the way people interact with technology, streamlining processes and improving productivity. As artificial intelligence continues to advance, the development of autonomous robots represents the next frontier in technological innovation. Autonomous robots are equipped with sensors, processors, and algorithms that enable them to operate independently without human intervention. Examples of autonomous robots include self-driving cars, cleaning robots, and robotic arms used in manufacturing. These robots demonstrate the potential for AI to enhance efficiency and safety by performing tasks that are tedious, dangerous, or time-consuming for humans. The development of autonomous robots also poses ethical and technical challenges, such as ensuring the safety and accountability of these machines in complex environments. The integration of AI in robotics holds promise for transforming industries such as transportation, healthcare, and agriculture, offering new opportunities for efficiency and innovation. The real-world applications of artificial intelligence, particularly autonomous robots, span a wide range of sectors, including medicine, logistics, agriculture, and security. In medicine, robots are being used to assist surgeons in delicate procedures, while in agriculture, drones equipped with AI algorithms are optimizing crop management and harvests. Autonomous robots in logistics are streamlining warehouse operations and delivery services, while in security, robots are enhancing surveillance and threat detection capabilities. While the benefits of AI-driven automation are vast, there are also risks associated with the potential loss of jobs and privacy concerns. As society navigates the future of artificial intelligence, a balance must be struck between embracing technological innovation and addressing the ethical implications to ensure that AI benefits humanity as a whole.

IV. NEURAL NETWORKS

Neural networks have become a critical component in the evolution of artificial intelligence, enabling machines to learn from data and make decisions without explicit programming. These complex systems of interconnected nodes, inspired by the structure of the human brain, have revolutionized various fields such as image and speech recognition, natural language processing, and autonomous decision-making. By simulating the way neurons communicate with each other, neural networks are able to analyze massive amounts of data and recognize patterns that would be impossible for humans to process manually. This ability has significantly improved the accuracy and efficiency of AI systems, leading to breakthroughs in diverse applications from healthcare diagnostics to self-driving cars. Through the integration of machine learning techniques and neural networks, AI-powered technologies like virtual assistants have evolved from simple voice-activated tools to sophisticated systems capable of understanding and responding to complex commands in natural language. This advancement has not only enhanced user experience but also enabled virtual assistants to perform a wider range of tasks, such as providing personalized recommendations, managing schedules, and even conducting basic conversations. The development of neural networks has played a crucial role in enabling these virtual assistants to continuously improve their performance through feedback loops and adaptive learning algorithms, making them indispensable tools in our daily lives. As we look to the future of artificial intelligence, autonomous robots stand out as the next frontier in advancing human-robot interaction and automation. These intelligent machines, equipped with sensors, actuators, and neural network algorithms, have the potential to revolutionize industries ranging from manufacturing and logistics to healthcare and agriculture. Autonomous robots have the capability to perform complex tasks with precision and efficiency, minimizing errors and increasing productivity. The widespread adoption of autonomous robots also raises ethical concerns related to job displacement, data privacy, and safety. As we continue to push the boundaries of AI technology, it is crucial to address these challenges and ensure that the benefits of autonomous robots are leveraged for the greater good of society.

Overview of Neural Networks

With the advancement of natural language processing technologies, virtual assistants have evolved from simply recognizing speech to understanding and responding to complex commands with greater accuracy and efficiency. This progress has been driven by the integration of machine learning algorithms and neural network models into the design and operation of virtual assistants. These sophisticated technologies have enabled virtual assistants to not only recognize words and phrases but also interpret the context, analyze semantics, and generate appropriate responses, leading to a more natural and intuitive interaction with users. Autonomous robots represent the latest frontier in artificial intelligence, defined by their ability to operate independently, make decisions, and adapt to changing environments without human intervention. Examples of current autonomous robots include self-driving vehicles, robotic cleaners, and industrial robots used in manufacturing and logistics. These machines leverage advanced sensors, machine learning algorithms, and neural networks to perceive their surroundings, navigate complex environments, and carry out specific tasks efficiently. The development of autonomous robots also presents significant technological and ethical challenges, including issues related to safety, security, privacy, and the potential impact on employment in various industries. Use Cases and Real-World Applications. The applications of autonomous robots span across diverse sectors, including medicine, logistics, agriculture, and security, where these machines have the potential to revolutionize operational workflows and enhance overall productivity. In the medical field, autonomous robots can assist in surgery, patient care, and drug delivery, while in agriculture, they can optimize crop management and harvesting processes. The widespread adoption of autonomous robots also raises concerns about potential risks and ethical implications, such as job displacement, data privacy, and the need for regulatory frameworks to ensure safe and responsible deployment of these technologies in society.

Deep Learning and Neural Networks

As deep learning and neural networks continue to advance, the potential for artificial intelligence to revolutionize various industries becomes increasingly apparent. Through the integration of machine learning and neural network technologies, virtual assistants have evolved from simple speech recognition tools to sophisticated systems capable of natural language understanding. This progression has significantly enhanced the capabilities of virtual assistants, enabling them to interpret complex commands and provide more accurate and relevant responses. The ability of these AI-powered systems to learn and improve over time through neural networks has transformed the way individuals interact with technology, paving the way for more personalized and efficient user experiences. The next frontier of artificial intelligence lies in the development of autonomous robots. These sophisticated machines, equipped with advanced sensors and decision-making capabilities, have the potential to revolutionize various sectors, from transportation to manufacturing. Autonomous robots can perform tasks in dynamic and unpredictable environments, showcasing the power of deep learning algorithms and neural networks in enabling autonomy and adaptability. The widespread adoption of autonomous robots also poses significant technological and ethical challenges, such as ensuring the safety and security of these systems and addressing concerns about job displacement and societal impacts. The evolution of artificial intelligence from virtual assistants to autonomous robots signifies a paradigm shift in the field of technology. As deep learning and neural networks continue to drive innovation in AI, the possibilities for leveraging these technologies to improve efficiency, productivity, and overall quality of life are endless. It is essential to approach the development and deployment of AI systems with caution, considering the ethical implications and potential risks associated with their widespread adoption. By striking a balance between technological innovation and ethical considerations, society can harness the full potential of artificial intelligence while ensuring that these advancements benefit humanity as a whole.

Neural Networks in Autonomous Robots

The integration of neural networks in autonomous robots marks a significant advancement in the field of artificial intelligence. Neural networks, inspired by the structure of the human brain, have the ability to process complex data and learn from it, enabling robots to make decisions autonomously. These networks, through deep learning algorithms, can analyze vast amounts of data and adapt their behavior based on new inputs, allowing for more sophisticated and refined decision-making processes in autonomous robots. By leveraging neural networks, autonomous robots can exhibit higher levels of cognitive capabilities, such as perception, reasoning, and decision-making. In the realm of autonomous vehicles, neural networks enable the vehicle to detect and classify objects in real-time, predict the behavior of other road users, and make split-second decisions to ensure safe navigation. This level of intelligence in robots has the potential to revolutionize industries such as transportation, manufacturing, and healthcare, where precise decision-making and adaptive behavior are crucial for operational efficiency. The incorporation of neural networks in autonomous robots also raises ethical and technological challenges. The reliance on advanced AI technology, powered by neural networks, brings concerns regarding data privacy, cybersecurity, and the potential impact of autonomous robots on the workforce. As these robots become more autonomous and capable of performing complex tasks, questions arise about the accountability and transparency of their decision-making processes. It is essential for policymakers, researchers, and industry stakeholders to address these challenges proactively to ensure the responsible and ethical deployment of autonomous robots in society.

V. NATURAL LANGUAGE PROCESSING

Advances in natural language processing have been at the forefront of the evolution of artificial intelligence, enabling virtual assistants to understand and respond to human language with increasing accuracy and sophistication. From basic speech recognition capabilities to complex natural language understanding, NLP technologies have revolutionized the interactions between users and virtual assistants like Siri, Alexa, and Google Assistant. By leveraging machine learning algorithms and neural network technologies, these virtual assistants can now interpret and process complex commands, leading to more personalized and intuitive user experiences. The evolution of NLP has played a pivotal role in enhancing the overall functionality and effectiveness of virtual assistants, making them indispensable tools in various industries and everyday life. As we move towards the next frontier of artificial intelligence, autonomous robots are poised to become the next game-changer in technology. Autonomous robots, equipped with advanced sensors, artificial intelligence algorithms, and decision-making capabilities, have the potential to revolutionize industries such as transportation, manufacturing, healthcare, and agriculture. These robots can perform tasks autonomously without human intervention, improving efficiency, productivity, and safety in various domains. The development of autonomous robots also poses significant technological and ethical challenges, such as ensuring the safety of human interactions, addressing concerns about job displacement, and navigating the regulatory landscape surrounding autonomous systems. The integration of autonomous robots into various sectors has already paved the way for transformative applications and capabilities. From autonomous vehicles navigating city streets to cleaning robots maintaining hygiene in hospitals, autonomous robots are redefining the possibilities of automation and robotics. These technologies have the potential to streamline operations, reduce costs, and improve the quality of services in diverse fields. The widespread adoption of autonomous robots also raises important questions about the ethical implications, societal impact, and potential risks associated with their deployment. As we navigate the evolving landscape of artificial intelligence and robotics, it is essential to consider the ethical dimensions and regulatory frameworks that will shape the future of autonomous systems.

Definition and Importance

As artificial intelligence continues to advance, the definition of autonomous robots becomes increasingly relevant. Autonomous robots can be defined as machines that can perform tasks without human intervention, relying on sensors, algorithms, and decision-making capabilities. These robots have the ability to adapt to new situations, learn from their environment, and make decisions based on the information they gather. Their importance lies in their potential to revolutionize various industries, from manufacturing to healthcare, by increasing efficiency, reducing costs, and improving safety. The significance of autonomous robots can be seen in their real-world applications across different sectors. Autonomous vehicles are transforming the transportation industry by offering safer and more efficient modes of travel. In the healthcare sector, surgical robots are enhancing precision and reducing the risk of human error during surgeries. In agriculture, autonomous robots are revolutionizing farming practices by increasing crop yields and reducing the need for manual labor. These examples highlight the potential of autonomous robots to streamline processes, increase productivity, and ultimately improve the quality of our lives. The development of autonomous robots also brings about technological and ethical challenges that must be addressed. Technological challenges include ensuring the reliability and safety of these robots, as well as addressing issues related to connectivity and data security. Ethical considerations involve questions regarding the impact of autonomous robots on employment, the responsibility for accidents involving autonomous systems, and the potential misuse of AI technologies. As we navigate the evolution of artificial intelligence towards autonomous robots, it is crucial to carefully consider these challenges and work towards creating a regulatory framework that promotes innovation while safeguarding against potential risks.

NLP Techniques

Advances in natural language processing techniques have played a crucial role in enhancing the capabilities of virtual assistants, enabling them to better understand and respond to human language. NLP algorithms have evolved significantly, allowing virtual assistants like Siri, Alexa, and Google Assistant to interpret and process complex commands with greater accuracy. By integrating machine learning and neural network technologies, these virtual assistants can now provide more personalized and contextually relevant responses to user queries. As a result, the integration of NLP techniques has significantly improved the overall user experience and functionality of virtual assistants, making them more efficient and effective in assisting users in various tasks. The continuous development and refinement of NLP techniques have paved the way for the next frontier of artificial intelligence: autonomous robots. These advanced robots are equipped with sophisticated algorithms and sensors that enable them to perform tasks independently without human intervention. Autonomous vehicles, cleaning robots, and robots in the industrial sector are just a few examples of how NLP techniques have been instrumental in enabling robots to understand and respond to their environment in real-time. The development of autonomous robots also poses technological and ethical challenges, such as ensuring the safety and security of these machines and addressing concerns about potential job displacement. The widespread adoption of autonomous robots across various sectors, including medicine, logistics, agriculture, and security, highlights the transformative impact of NLP techniques on business efficiency and productivity. While the use of autonomous robots offers numerous benefits, such as increased speed and accuracy in completing tasks, it also raises concerns about the potential risks associated with relying heavily on AI-driven machines. As we look ahead to the future of artificial intelligence, it is essential to consider the ethical implications of integrating NLP techniques into autonomous systems and to strike a balance between technological innovation and social responsibility to ensure that AI continues to benefit society as a whole.

NLP in Conversational AI

The advancements in natural language processing have been instrumental in shaping the capabilities of conversational AI. NLP allows virtual assistants to not only transcribe speech into text but also comprehend the meaning behind the words, enabling more natural interactions between humans and machines. Through the integration of machine learning and neural network technologies, virtual assistants have become more adept at understanding context, identifying sentiment, and providing relevant responses. This evolution in NLP has significantly enhanced the user experience by making interactions with AI-driven systems more intuitive and seamless. With the progression of NLP, virtual assistants have gained the ability to process complex commands and follow conversational threads, mimicking the fluidity of human communication. By leveraging sophisticated algorithms and large datasets, these AI systems can offer personalized recommendations, answer queries with greater accuracy, and anticipate user needs based on previous interactions. The refinement of NLP has paved the way for virtual assistants to evolve beyond simple question-answer interfaces into intelligent conversational agents capable of engaging in more sophisticated dialogues. As a result, NLP has been a crucial component in bridging the gap between humans and machines, transforming how we interact with technology in our daily lives. In the realm of conversational AI, the continued advancements in NLP hold the key to unlocking new possibilities for autonomous robots. As autonomous robots rely on seamless communication to navigate their environment, understand commands, and interact with humans, the integration of advanced NLP capabilities is vital for their development. By enhancing the linguistic and cognitive abilities of autonomous systems, NLP plays a pivotal role in enabling robots to carry out complex tasks autonomously and collaborate effectively with human operators. The convergence of NLP with robotics signifies a significant leap forward in the

evolution of artificial intelligence, with the potential to revolutionize industries ranging from manufacturing and logistics to
healthcare and service sectors.

VI. ROBOTICS

Technological advancements in robotics have revolutionized various industries, enabling the development of autonomous robots that can perform tasks with minimal human intervention. These robots, equipped with sophisticated sensors and algorithms, have the ability to perceive their environment, make decisions, and execute actions without constant human oversight. From autonomous vehicles navigating city streets to robots in warehouses optimizing logistics operations, the potential applications of autonomous robots are vast and continue to expand. The development of autonomous robots also poses significant challenges, such as ethical considerations regarding decision-making processes and potential job displacement as automation becomes more prevalent in the workforce. One of the key characteristics of autonomous robots is their ability to operate independently, using sensors and artificial intelligence algorithms to interact with their surroundings and perform tasks without direct human control. This level of autonomy opens up new possibilities for improving efficiency, precision, and safety in various fields. Autonomous drones can be used for aerial surveying in agriculture, monitoring wildlife populations, or delivering goods to remote areas. In manufacturing, robots equipped with artificial intelligence can collaborate with human workers to optimize production processes and increase overall productivity. As these technologies continue to evolve, it is crucial to address safety concerns and ethical implications to ensure that autonomous robots are developed and deployed responsibly. The evolution of artificial intelligence from virtual assistants to autonomous robots represents a significant milestone in the field of robotics. The integration of advanced technologies such as machine learning, natural language processing, and computer vision has enabled the development of robots that can operate independently and adapt to changing environments. While the potential benefits of autonomous robots are substantial, it is essential to address ethical, societal, and regulatory considerations to ensure the responsible use of these technologies. As autonomous robots become more prevalent in various sectors, it is vital to strike a balance between innovation and ethical considerations to harness the full potential of artificial intelligence for the betterment of society.

Introduction to Robotics

As artificial intelligence continues to advance, one of the key milestones in its evolution has been the development of robotics. Robotics represents a convergence of various technologies, including AI, machine learning, and sensor technologies, to create machines capable of performing tasks autonomously. These robots are designed to emulate human actions and interactions, with the ultimate goal of improving efficiency and productivity in various industries. The field of robotics is broad, encompassing a wide range of applications, from industrial robots used in manufacturing processes to autonomous vehicles navigating our roads. The transition from virtual assistants to autonomous robots represents a significant shift in the role of AI in our daily lives. Virtual assistants, such as Siri, Alexa, and Google Assistant, have become ubiquitous, providing users with a convenient way to access information and perform various tasks through voice commands. These early AI applications laid the groundwork for more advanced systems capable of natural language understanding and context awareness. As technology progresses, we are now seeing the emergence of autonomous robots that can operate independently in dynamic environments, making decisions and adapting to changing circumstances in real-time. The development of autonomous robots raises important ethical and societal considerations. As these machines become more sophisticated and integrated into our daily lives, questions arise about the impact on employment, security, and privacy. Ensuring that autonomous robots are designed and used ethically and responsibly is crucial to harnessing their potential benefits while mitigating potential risks. As we navigate this rapidly evolving landscape of artificial intelligence and robotics, it is imperative to strike a balance between technological innovation and ethical considerations to ensure that these advancements benefit society as a whole.

Types of Robots

The development of autonomous robots represents the next frontier in artificial intelligence, offering a glimpse into a future where machines can operate independently without human intervention. These robots are equipped with advanced sensors, algorithms, and decision-making capabilities that enable them to navigate and interact with their environment autonomously. Examples of current autonomous robots include self-driving cars, cleaning robots, and industrial robots that can perform complex tasks with precision. The development of autonomous robots also poses significant technological and ethical challenges, including concerns about safety, privacy, and the potential impact on the future of work. The use cases and real-world applications of autonomous robots span across various sectors, revolutionizing industries such as medicine, logistics, agriculture, and security. In medicine, autonomous robots can assist in surgeries, deliver medications, and provide support to healthcare professionals. In logistics, these robots can optimize supply chain operations, warehouse management, and last-mile delivery. While the adoption of autonomous robots promises increased efficiency and productivity, it also raises concerns about job displacement, ethical implications, and the need for regulations to ensure responsible use of AI technologies. As the evolution of artificial intelligence continues to unfold, the future of autonomous robots holds great promise for society. Emerging trends in AI and robotics point towards further advancements in autonomy, adaptability, and intelligence in machines. These developments also demand a careful reflection on the ethical dilemmas and societal implications of integrating autonomous robots into our daily lives. Balancing technological innovation with ethical considerations will be crucial in shaping a future where AI benefits humanity while addressing the challenges posed by the proliferation of autonomous robots.

Integration of AI in Robotics

In the realm of robotics, the integration of artificial intelligence has marked a significant milestone, paving the way for the development of autonomous robots. These robots, equipped with advanced AI capabilities, are capable of making independent decisions and executing tasks without human intervention. The integration of AI in robotics has expanded the horizons of what robots can achieve, enhancing their adaptability and problem-solving abilities. By leveraging AI algorithms and machine learning techniques, autonomous robots can navigate complex environments, learn from experience, and interact with their surroundings in a more sophisticated manner. One key aspect of the integration of AI in robotics is the development of autonomous vehicles, which have the potential to revolutionize transportation systems and enhance road safety. These vehicles utilize AI technologies such as computer vision, sensor fusion, and deep learning to perceive their environment, plan their routes, and make real-time decisions while on the road. The advent of autonomous vehicles represents a new era in mobility, with implications for urban planning, sustainability, and the automotive industry as a whole. As these technologies mature, the prospect of fully autonomous vehicles navigating city streets alongside human drivers is becoming increasingly tangible. The integration of AI in robotics extends beyond autonomous vehicles to other domains such as healthcare, manufacturing, and agriculture. In the medical field, AI-powered robots can assist surgeons during complex procedures, analyze medical images with high precision, and provide support in patient care. In manufacturing, collaborative robots equipped with AI capabilities can work alongside human operators, enhancing productivity and efficiency on the factory floor. In agriculture, autonomous drones and robots can optimize crop management, monitor soil conditions, and improve yields. The widespread adoption of AI-powered robots in these sectors is reshaping traditional workflows, increasing operational efficiency, and unlocking new opportunities for innovation.

VII. ETHICS IN AI DEVELOPMENT

Ethics in AI development is a crucial aspect that needs to be carefully considered as technology continues to advance. One of the primary ethical concerns in AI development is the issue of bias. AI systems are trained on vast amounts of data, and if this data is biased or flawed, it can lead to biased outcomes. Facial recognition software has been found to have higher error rates for people of color, highlighting the bias that exists in these systems. Engineers and developers need to be mindful of these biases and work towards creating more inclusive and equitable AI technologies. Another ethical consideration in AI development is the potential impact on privacy and data security. As AI systems collect and process massive amounts of personal data, there is a risk of this information being misused or compromised. It is essential for developers to implement robust security measures and data protection protocols to safeguard user information. There is a need for transparency in how AI systems operate and make decisions, to ensure that users understand how their data is being used and have control over their information. The ethical implications of autonomous robots raise questions about accountability and decision-making. As these robots become more autonomous and independent in their actions, the responsibility for any errors or harm caused by the robots becomes unclear. Establishing clear guidelines and regulations for the development and deployment of autonomous robots is crucial to ensure that ethical standards are upheld. Incorporating ethical principles such as transparency, fairness, and accountability into the design and programming of these robots is essential to mitigate potential ethical risks. The ethical considerations in AI development must be carefully navigated to ensure that technology serves the greater good and upholds ethical standards.

Importance of Ethical Considerations

As technology continues to advance at a rapid pace, the importance of ethical considerations in the development and implementation of artificial intelligence cannot be overstated. With the transition from virtual assistants to autonomous robots, the ethical implications become even more complex and significant. It is crucial for designers, engineers, and policymakers to carefully consider the ethical implications of AI, ensuring that these technologies are used in a way that respects human rights, promotes fairness, and prioritizes societal well-being. Ethical considerations play a fundamental role in ensuring that AI is developed and deployed responsibly, protecting individuals and communities from potential harm and misuse. One of the key reasons why ethical considerations are essential in the evolution of AI is the potential impact on job displacement and economic inequality. As autonomous robots become more prevalent in various industries, there is a growing concern about the displacement of human workers and the widening gap between the wealthy and the less fortunate. Ethical frameworks can help guide decision-making processes to ensure that the benefits of AI are equitably distributed and that measures are in place to support those affected by technological disruptions. By addressing these ethical challenges proactively, society can work towards creating a more inclusive and sustainable future where AI enhances human capabilities rather than replacing them. Ethical considerations also play a critical role in safeguarding privacy, data security, and the protection of vulnerable populations. As autonomous robots collect and process vast amounts of data, there is a heightened risk of privacy breaches and misuse of personal information. Ethical guidelines can help establish clear boundaries and standards for data handling, ensuring that individuals have control over their data and that data usage is conducted in a transparent and responsible manner. Ethical considerations can help mitigate potential risks associated with bias, discrimination, and unintended consequences in AI systems, fostering trust and accountability in the design and deployment of these technologies. Integrating ethical principles into the development of AI is essential for creating a future where technology serves the common good and upholds the values of justice, equity, and human dignity.

Ethical Issues in AI

As artificial intelligence continues to advance, ethical issues surrounding its development and deployment have become more pressing. One key ethical concern in AI is the potential for biased algorithms, which can perpetuate discrimination and inequality. This bias can be unintentionally embedded in the data used to train AI systems, leading to biased outcomes in decision-making processes. In recruitment algorithms, biases in the training data can result in discriminatory hiring practices that disadvantage certain groups. Addressing these biases requires careful oversight and monitoring of AI systems to ensure fairness and equity in their operations. Another ethical issue in AI is the issue of privacy and data security. AI systems often rely on vast amounts of personal data to operate effectively, raising concerns about how this data is collected, stored, and used. Unauthorized access to sensitive personal information can lead to privacy breaches and potential harm to individuals. The use of AI in surveillance and monitoring activities can infringe on people's right to privacy and raise concerns about mass surveillance. Striking a balance between the benefits of AI innovation and the protection of individual privacy rights is crucial for ensuring the ethical development and deployment of AI technologies. Concerns about the impact of AI on employment and the economy have also come to the forefront of ethical discussions. The automation of jobs by AI systems can lead to widespread unemployment and economic disruption, particularly in industries heavily reliant on manual labor. This raises questions about the responsibility of companies and governments to mitigate the negative effects of automation on workers through retraining programs and social safety nets. Ethical considerations must also be taken into account when developing AI technologies that have the potential to reshape the labor market and disrupt traditional economic structures. Navigating these ethical challenges in AI requires a comprehensive and transparent approach that prioritizes the well-being and rights of individuals in the face of technological advancements.

Regulatory Frameworks
With the rapid advancement of artificial intelligence technologies, regulatory frameworks have become essential to ensure their responsible development and deployment. These frameworks serve as guidelines that govern the use of AI systems, covering aspects such as data privacy, algorithmic transparency, accountability, and safety. As AI applications expand into

various domains, ranging from healthcare to finance, regulatory bodies worldwide are facing the challenge of keeping pace with the evolving technology landscape while safeguarding the rights and interests of individuals and society as a whole. One key aspect of regulatory frameworks for AI is the establishment of ethical guidelines to govern the design and deployment of intelligent systems. Ethical considerations play a crucial role in ensuring that AI technologies are developed in a manner that upholds fundamental human values and principles. Issues such as transparency, fairness, accountability, and bias mitigation are central to the ethical framework for AI. By embedding ethical principles into regulatory guidelines, policymakers can promote the responsible development of AI and address concerns related to the potential risks and impact of intelligent systems on individuals and society. In addition to ethical considerations, regulatory frameworks for AI also focus on addressing legal and compliance challenges associated with the use of intelligent systems. From data protection regulations such as the General Data Protection Regulation (GDPR) to specific guidelines on AI deployment in sensitive sectors like healthcare and finance, regulatory bodies are working to establish clear rules and standards for AI applications. By creating a coherent regulatory environment, policymakers aim to facilitate the integration of AI technologies into existing legal frameworks while minimizing potential risks and ensuring accountability for AI-powered decisions and actions.

VIII. CHALLENGES IN AI ADVANCEMENT

As artificial intelligence continues to advance, there are several challenges that researchers and developers face in pushing the boundaries of this technology. One of the key challenges lies in the ethical implications of AI, particularly in the realm of autonomous robots. As these robots become increasingly autonomous and capable of making decisions without human intervention, questions arise about the ethical principles that should guide their behavior. Issues such as accountability, transparency, and bias in AI algorithms need to be carefully considered to ensure that autonomous robots operate in a way that aligns with societal values and norms. Another significant challenge in the advancement of AI is the need for robust and reliable data. AI systems rely heavily on vast amounts of data to learn and make decisions. Ensuring the quality, accuracy, and fairness of the data used to train these systems can be a complex task. Biases in data sets can lead to skewed outcomes and hinder the effectiveness of AI applications. Researchers must address these challenges by implementing rigorous data collection and processing procedures to minimize bias and ensure that AI systems produce fair and unbiased results. The rapid pace of technological innovation poses a challenge in keeping AI systems up-to-date and secure. As AI technologies evolve, maintaining the security and integrity of AI systems becomes increasingly crucial. Vulnerabilities in AI systems can be exploited by malicious actors, leading to potential harm and disruption. This requires continuous monitoring, updating, and testing of AI systems to ensure that they remain resilient to cyber threats. Overcoming these challenges will be essential in advancing AI technology responsibly and sustainably in the future.

Limitations of Current AI Systems

The rapid advancements in artificial intelligence have undoubtedly revolutionized various aspects of our lives. It is essential to recognize the limitations of current AI systems to understand where further improvements are needed. One significant limitation is the lack of common sense reasoning, which can lead AI systems to make irrational or erroneous decisions when faced with unfamiliar situations. While AI has excelled in specific tasks that involve pattern recognition and data processing, its ability to handle complex reasoning and abstract concepts still falls short of human intelligence. This limitation poses challenges in applications where nuanced understanding or creative problem-solving is required. Another limitation of current AI systems is their susceptibility to bias and discrimination. AI algorithms are trained on historical data, which may contain biased information reflecting societal prejudices. As a result, AI systems can perpetuate and even exacerbate existing biases when making decisions, particularly in sensitive areas such as hiring practices or criminal justice. To address this issue, researchers are exploring ways to mitigate bias in AI algorithms, such as through data preprocessing and algorithmic transparency. This remains a key challenge that must be addressed to ensure fair and ethical AI applications. The lack of explainability and transparency in AI decision-making processes poses a significant limitation. Complex deep learning algorithms, for example, operate as black boxes, making it challenging to understand how and why a particular decision was made. In critical applications like healthcare or autonomous vehicles, where human lives are at stake, the inability to interpret AI decisions can hinder trust and adoption. As AI systems become more prevalent in society, there is a growing need for explainable AI that can provide insights into the decision-making process, increasing transparency and fostering trust among users and stakeholders.

Bias and Fairness in AI

Advancements in artificial intelligence have brought about a multitude of benefits in various aspects of human life, from virtual assistants to autonomous robots. With these technological strides come inherent biases that can affect the fairness and ethicality of AI systems. Bias in AI can stem from a variety of sources, including data collection methods, algorithmic design, and the lack of diversity in the teams developing these systems. Biased data used to train AI models can lead to discriminatory outcomes, such as facial recognition systems that are less accurate for certain demographic groups. As AI continues to permeate different industries and aspects of society, addressing bias becomes paramount to ensure fairness and equity in the deployment of these technologies. Efforts to mitigate bias in AI systems include the development of fairness-aware algorithms that aim to reduce discrimination and promote transparency in decision-making processes. By incorporating fairness metrics into the design and evaluation of AI models, researchers and engineers can identify and rectify biases before deploying these systems in real-world applications. Fostering diversity and inclusivity within AI teams can help mitigate unconscious biases and promote the development of impartial and equitable technologies. Collaborative efforts between industry, academia, and policymakers are essential to establish guidelines and regulations that ensure the ethical and fair deployment of AI systems. While bias in AI remains a significant challenge, addressing these issues is crucial to ensure that AI technologies benefit society as a whole. Ethical considerations must be integrated into the design and implementation of AI systems to uphold principles of fairness, transparency, and accountability. By fostering a culture of ethical AI development and responsible innovation, we can harness the transformative potential of artificial intelligence while mitigating the risks associated with bias and discrimination. As the evolution of AI continues, it is imperative to prioritize fairness and equity to build a future where technology serves the collective good.

Security and Privacy Concerns

As artificial intelligence continues to advance, security and privacy concerns have become increasingly prevalent. The integration of AI into various aspects of our lives, from virtual assistants to autonomous robots, raises questions about the protection of sensitive data and the potential for misuse. One major issue is the vulnerability of AI systems to cyber-attacks, which could compromise personal information or even lead to malicious activities. As these technologies become more sophisticated, the need for robust cybersecurity measures becomes paramount to safeguard against potential threats and ensure the integrity of AI systems. The collection and storage of massive amounts of data by AI systems also raise privacy concerns. Virtual assistants, for example, often require access to personal information to provide personalized responses and services. This data collection raises questions about the extent to which individuals' privacy is being violated and the potential for unauthorized access to sensitive data. Striking a balance between the benefits of AI-driven services and the protection of user privacy is crucial to maintain trust in these technologies and ensure ethical practices in their development and implementation. The deployment of autonomous robots in various industries introduces new challenges in terms of security and privacy. These robots, equipped with advanced sensors and AI capabilities, have the potential to collect and analyze vast amounts of data in real-time. While this data can be used to enhance efficiency and decision-making, it also raises concerns about who has access to this information and how it is being used. Regulations and guidelines for data protection and privacy must evolve alongside the development of autonomous robots to address these issues and ensure that the benefits of AI technology are realized without compromising security and privacy.

IX. FUTURE OF AI

The rapid advancement of artificial intelligence in recent years has sparked numerous debates about the future of this transformative technology. One significant development that has captured the imagination of researchers and the public alike is the evolution from virtual assistants to autonomous robots. Virtual assistants, like Siri, Alexa, and Google Assistant, have been instrumental in familiarizing society with AI technology by providing personalized assistance and performing a range of tasks. These virtual assistants have demonstrated the potential of AI to enhance everyday life and streamline processes in various industries, laying the foundation for more sophisticated applications. As technology continues to progress, the focus has shifted towards achieving natural language understanding in AI systems. This advancement involves enhancing the ability of virtual assistants to comprehend and respond to complex commands, making them more intuitive and user-friendly. The integration of machine learning and neural network technologies has played a crucial role in improving the accuracy and efficiency of virtual assistants, paving the way for more sophisticated applications in the future. By mastering natural language understanding, AI systems are becoming increasingly adept at interpreting human language and carrying out tasks with greater precision and sophistication. Moving forward, the advent of autonomous robots represents the next frontier in AI development. These robots, equipped with the ability to operate independently and make decisions based on their environment, hold immense potential for revolutionizing various industries. From autonomous vehicles to cleaning robots and industrial automation, these machines are set to disrupt traditional paradigms and reshape the way tasks are performed. The development of autonomous robots also poses significant technological and ethical challenges that must be addressed to ensure their safe and responsible integration into society. As AI technology continues to evolve, it is essential to consider the implications of these advancements on ethical standards, regulatory frameworks, and societal well-being.

Emerging Trends in AI

Advances in artificial intelligence have paved the way for emerging trends that are reshaping the technology landscape. One such trend is the shift from virtual assistants to autonomous robots, marking a significant evolution in the capabilities of AI systems. Virtual assistants, such as Siri, Alexa, and Google Assistant, were among the first applications of artificial intelligence in everyday life. These systems initially focused on basic tasks like retrieving information or setting reminders but have since advanced to understand and respond to complex commands, thanks to improvements in natural language processing and the integration of machine learning technologies. As technology continues to progress, the focus has shifted towards autonomous robots as the next frontier of artificial intelligence. These robots are capable of operating independently, making decisions based on their surroundings and tasks. Current examples include autonomous vehicles, cleaning robots, and robots used in industrial settings. The development of autonomous robots poses technological and ethical challenges, as their widespread adoption raises concerns about safety, job displacement, and privacy issues. Understanding these challenges is crucial in ensuring that autonomous robots are integrated responsibly into society. The use cases and real-world applications of autonomous robots span across various sectors, including medicine, logistics, agriculture, and security. These robots have the potential to revolutionize business operations, increasing efficiency and productivity. While the benefits are plentiful, the risks associated with the widespread adoption of autonomous robots cannot be overlooked. As AI and robotics continue to shape the future, it is essential to consider the implications of these technologies on society at large. By staying informed about emerging trends and actively participating in discussions on ethical considerations, we can effectively navigate the evolving landscape of artificial intelligence.

Potential Impact on Society

Advancements in artificial intelligence have the potential to significantly impact society in various ways. One major area where this impact is already evident is in the realm of virtual assistants. These AI-powered tools, such as Siri, Alexa, and Google Assistant, have become integral parts of everyday life for many people. They have transformed how we interact with technology, enabling us to complete tasks, find information, and control our devices through simple voice commands. Virtual assistants have also had a significant impact on various industries, from customer service to healthcare, by improving efficiency and streamlining processes. As technology continues to progress, we are witnessing a shift towards more advanced forms of AI, such as autonomous robots. These devices are capable of performing tasks and making decisions without human intervention, opening up new possibilities in fields like transportation, manufacturing, and healthcare. The deployment of autonomous robots, such as autonomous vehicles and cleaning robots, has the potential to revolutionize industries and improve productivity. The development of these robots also raises important ethical questions and challenges, such as concerns about job displacement and the potential for misuse of AI technology. Looking ahead, the widespread adoption of autonomous robots and other AI applications will have far-reaching implications for society. On one hand, these technologies have the potential to bring about significant benefits, including increased efficiency, improved safety, and enhanced quality of life. On the other hand, there are risks to consider, such as the potential for bias in AI decision-making and concerns about data privacy. As we navigate the evolving landscape of artificial intelligence, it is crucial to strike a balance between technological innovation and ethical considerations to ensure that these advancements serve the greater good and benefit society as a whole.

Ethical and Regulatory Implications

As artificial intelligence progresses from virtual assistants to autonomous robots, a myriad of ethical and regulatory implications come to the forefront. The increasing autonomy and decision-making capabilities of AI systems raise concerns about accountability and liability in case of errors or accidents. The potential displacement of human workers by autonomous robots in various industries necessitates a reevaluation of labor laws and social safety nets to ensure a smooth transition to an automated workforce. These ethical dilemmas highlight the need for clear guidelines and regulations to govern the development and deployment of AI technologies, balancing innovation with ethical considerations. The issue of bias and fairness in AI algorithms poses a significant ethical challenge in the evolution towards autonomous robots. Research has shown that AI systems can inherit and perpetuate human biases present in the data they are trained on, leading to discriminatory outcomes in decision-making processes. Addressing these biases requires a combination of technical solutions, such as algorithmic transparency and accountability mechanisms, as well as ethical frameworks that promote diversity and inclusion in AI development. Regulatory bodies must play a crucial role in ensuring that AI systems are developed in a transparent and accountable manner to mitigate the risks of bias and discrimination. The evolution of AI from virtual assistants to autonomous robots underscores the urgent need for robust ethical frameworks and regulations to guide the responsible deployment of AI technologies. As AI continues to permeate various sectors of society, from healthcare to transportation, policymakers must collaborate with industry stakeholders and ethicists to anticipate and address the ethical and regulatory challenges that arise. By fostering a culture of ethical innovation and prioritizing human values in the development of AI systems, we can harness the transformative potential of AI while safeguarding the well-being of individuals and society at large.

X. QUANTUM COMPUTING IN AI

Recent advances in quantum computing have sparked great interest in the field of artificial intelligence. Quantum computing has the potential to revolutionize AI by exponentially increasing computational power and enabling more complex calculations to be performed in a fraction of the time compared to classical computers. This significant enhancement in processing capabilities could lead to breakthroughs in various AI applications, from natural language processing to machine learning algorithms. Quantum computing could enable AI systems to tackle problems that are currently considered too complex or time-consuming, opening up new possibilities for innovation and technological advancement in the field. As quantum computing continues to evolve, researchers are exploring ways to integrate quantum principles into AI systems to enhance their performance and capabilities. Quantum AI algorithms have been developed to leverage the unique properties of quantum computing, such as superposition and entanglement, to optimize problem-solving and decision-making processes. By harnessing the power of quantum mechanics, AI systems could achieve unprecedented levels of efficiency and accuracy, paving the way for revolutionary advancements in autonomous robots, medical diagnosis, financial modeling, and other critical domains. The synergy between quantum computing and AI holds immense promise for solving some of the most challenging problems faced by humanity, propelling us into a new era of innovation and discovery. Despite the potential benefits of quantum computing in AI, there are significant challenges that must be addressed to fully realize its potential. The integration of quantum principles into AI systems requires specialized expertise and resources, making it a complex and technically demanding endeavor. The delicate nature of quantum information necessitates robust error-correction mechanisms to mitigate the impact of noise and decoherence on calculations. Ethical considerations surrounding the use of quantum AI also raise important questions about privacy, security, and the responsible development of technology. As quantum computing and AI converge, it is essential for researchers, policymakers, and industry leaders to collaborate effectively to navigate these challenges and unlock the transformative power of quantum AI for the benefit of society.

Introduction to Quantum Computing

The concept of quantum computing represents a significant leap forward in the field of artificial intelligence, promising unprecedented computational power and the ability to solve complex problems at an exponentially faster rate than traditional computers. Unlike classical computers that rely on bits to process information, quantum computers use quantum bits, or qubits, which can exist in multiple states simultaneously due to the principles of superposition and entanglement. This allows quantum computers to perform calculations in parallel, making them ideally suited for tasks such as optimization, cryptography, and simulations that are beyond the capabilities of classical computers. The potential applications of quantum computing extend beyond traditional computational tasks, offering opportunities to revolutionize fields such as machine learning, drug discovery, and materials science. Quantum algorithms have the capacity to process and analyze vast amounts of data more efficiently, leading to faster and more accurate predictions in various domains. Quantum machine learning models can leverage the power of quantum entanglement to enhance pattern recognition and decision-making processes, paving the way for more advanced AI systems with greater capabilities and intelligence. The development of quantum computing poses new challenges and opportunities for researchers and policymakers alike. As quantum technology continues to evolve, it is essential to address issues related to data security, algorithm optimization, and ethical considerations. Ensuring the responsible and ethical deployment of quantum computing technologies will be crucial in harnessing their full potential while mitigating potential risks. By exploring the principles of quantum computing and its implications for artificial intelligence, we can gain valuable insights into the future of technology and its transformative impact on society.

Quantum Machine Learning Algorithms

Quantum machine learning algorithms represent a cutting-edge approach that leverages the principles of quantum mechanics to enhance traditional machine learning techniques. By harnessing the power of quantum computing, these algorithms have the potential to revolutionize the field of artificial intelligence by exponentially increasing computational speed and efficiency. Quantum algorithms like the quantum support vector machine and quantum neural networks show promise in solving complex optimization problems and performing pattern recognition tasks at a scale that is currently unattainable with classical computing systems. This advancement opens up new frontiers in AI research, enabling the development of more sophisticated models capable of handling vast amounts of data with unmatched precision and accuracy. One of the key advantages of quantum machine learning algorithms lies in their ability to process and analyze massive datasets in parallel, leading to significant improvements in training times and prediction accuracy. By exploiting quantum superposition and entanglement, these algorithms can explore a myriad of possible solutions simultaneously, providing a quantum leap in computational efficiency. Quantum machine learning algorithms have the potential to address the limitations of classical machine learning models, particularly in scenarios where the curse of dimensionality hinders the performance of traditional algorithms. Through quantum parallelism and interference, these algorithms can navigate high-dimensional feature spaces more effectively, leading to more robust and reliable AI models. The implementation of quantum machine learning algorithms also presents several challenges, including the need for large-scale quantum hardware and sophisticated error correction techniques to mitigate the effects of quantum decoherence and noise. The design and optimization of quantum circuits for specific machine learning tasks require expertise in both quantum computing and AI, making it essential to foster interdisciplinary collaboration between researchers in these fields. Despite these challenges, the potential of quantum machine learning algorithms to revolutionize AI in the coming years is undeniable, paving the way for a new era of intelligent systems that can surpass the limitations of classical computing and drive innovation across various industries.

Quantum Neural Networks
Quantum neural networks represent an exciting frontier in the

field of artificial intelligence, combining the principles of quantum computing with the power of neural networks. These networks leverage the unique properties of quantum mechanics, such as superposition and entanglement, to perform complex computations at an exponentially faster rate than classical computers. By harnessing the parallel processing capabilities of quantum systems, quantum neural networks have the potential to revolutionize the way in which AI algorithms are developed and executed. One key advantage of quantum neural networks is their ability to handle exponentially growing amounts of data in a more efficient manner than classical neural networks. This is particularly beneficial in tasks that require processing large datasets, such as image recognition, natural language processing, and optimization problems. Quantum neural networks can explore multiple solutions simultaneously, providing a more comprehensive and rapid approach to problem-solving. The quantum nature of these networks allows for complex interactions among neurons, leading to the development of more sophisticated learning algorithms. Quantum neural networks have the potential to address some of the limitations of classical neural networks, such as the vanishing gradient problem and overfitting. By utilizing quantum properties such as quantum superposition and entanglement, these networks can optimize training processes and enhance the learning capabilities of AI systems. As researchers continue to explore the possibilities of quantum computing in the realm of artificial intelligence, quantum neural networks offer a promising avenue for developing more advanced and efficient AI algorithms that can tackle complex problems across various industries.

XI. EXPLAINABLE AI

Advances in artificial intelligence have led to the development of Explainable AI (XAI), a branch of AI that focuses on making the decisions and processes of AI systems more transparent and understandable to humans. XAI aims to bridge the gap between the complex algorithms used by AI systems and the need for human users to trust and interpret the decisions made by these systems. By providing explanations and justifications for the decisions made by AI models, XAI enhances accountability, fosters trust, and enables humans to intervene when necessary. This is crucial in critical applications such as healthcare, finance, and autonomous vehicles, where the consequences of AI decisions can have significant real-world impact. One of the key components of XAI is the development of interpretable machine learning models that can provide insights into their decision-making process. This includes techniques such as feature importance analysis, attention mechanisms, and decision trees that allow users to understand how a model arrives at a particular outcome. XAI methods enable users to explore the limitations and biases of AI systems, ultimately leading to more robust and fair decision-making processes. By improving the transparency and interpretability of AI models, XAI provides a foundation for building trust between humans and intelligent systems, paving the way for more widespread adoption and integration of AI technologies in various industries. XAI plays a crucial role in addressing ethical considerations related to AI systems, such as bias, accountability, and fairness. With the increasing reliance on AI systems in critical decision-making processes, it is essential to ensure that these systems are not only accurate but also ethical and transparent. By providing explanations for AI decisions, XAI allows for the identification and mitigation of biases that may be present in the data or algorithms used by AI systems. This transparency also holds AI developers and operators accountable for the decisions made by their systems, promoting responsible and ethical use of AI technologies in society. Explainable AI represents a vital step towards ensuring that AI systems are not only intelligent but also transparent, ethical, and accountable in their decision-making processes.

Importance of Explainable AI
The importance of explainable AI cannot be overstated, especially as we move towards the deployment of more advanced autonomous systems. While AI has made remarkable advancements in recent years, the lack of transparency in the decision-making processes of these systems poses significant risks in terms of accountability and reliability. Explainable AI plays a crucial role in ensuring that humans can understand and trust the reasoning behind the decisions made by AI systems. This transparency not only fosters user confidence but also helps in identifying and rectifying biases or errors in the algorithms, ultimately leading to more ethical and reliable AI applications. Explainable AI is essential for promoting ethical considerations in the development and deployment of autonomous robots. As these systems are designed to operate without constant human supervision, it becomes crucial to understand how and why they make certain decisions. By providing explanations for the actions taken by autonomous robots, developers and users can ensure that these machines adhere to ethical standards and are accountable for their behavior. This transparency also helps in

addressing concerns related to safety, privacy, and potential risks associated with autonomous systems, thereby paving the way for the responsible integration of AI in society. The adoption of explainable AI can have significant implications for the legal and regulatory frameworks governing AI technologies. As these systems become more sophisticated and autonomous, it becomes imperative to establish guidelines and standards for their development and use. Explainable AI can provide insights into the decision-making process of these systems, enabling regulators and policymakers to enforce laws and regulations effectively. By promoting transparency and accountability, explainable AI can help in addressing legal challenges such as liability, data protection, and privacy concerns, ensuring that AI technologies are deployed in a manner that aligns with societal values and norms.

Techniques for Interpretable AI Models

Advances in artificial intelligence have led to the development of complex and sophisticated models that can perform a wide range of tasks. One key challenge in AI research is ensuring that these models are interpretable, meaning that their decisions and processes can be understood by humans. Techniques for creating interpretable AI models have gained significant attention in recent years. One approach involves using simpler models, such as decision trees or linear regression, in place of more complex algorithms like deep neural networks. By using models that are easier to interpret, researchers and practitioners can better understand how AI systems arrive at their conclusions. Another technique for achieving interpretability in AI models is through the use of visualization tools. These tools can help researchers and users visualize the inner workings of a model, making it easier to understand how different variables contribute to the final output. Techniques like saliency maps, which highlight important features in an image or text, can provide valuable insights into how a model is making its decisions. By making use of visualizations, stakeholders can gain a deeper understanding of AI systems and potentially uncover biases or errors that may be present in the model. In addition to using simpler models and visualization tools, another effective technique for creating interpretable AI models is through the incorporation of feature importance techniques. These techniques allow researchers to identify which features or variables are most influential in driving the model's predictions. By understanding the importance of different factors, stakeholders can gain valuable insights into the decision-making process of AI models. This level of transparency is crucial for building trust in AI systems and ensuring that they are used responsibly in various domains. The development of techniques for interpretable AI models is essential for fostering trust, transparency, and accountability in the deployment of artificial intelligence technologies.

Applications in Critical Decision-Making Systems

Autonomous robots represent the next frontier in the evolution of artificial intelligence, showcasing the remarkable advancements in technology that have paved the way for these sophisticated systems. Defined by their ability to operate independently and make decisions based on their environment, autonomous robots hold immense potential for revolutionizing various industries. From autonomous vehicles navigating city streets to cleaning robots maintaining hygiene in homes and workplaces, the applications of these robots are vast and diverse. The development of autonomous robots also presents significant technological and ethical challenges that must be addressed, such as ensuring safety, preventing misuse, and addressing the potential impact on the workforce. One of the key factors driving the rise of autonomous robots is their ability to perform tasks that are considered too dangerous, mundane, or time-consuming for humans. In the medical field, autonomous surgical robots are being developed to assist surgeons in delicate procedures with precision and efficiency. In logistics, autonomous drones and robots are revolutionizing the warehousing and delivery processes, speeding up operations and reducing costs. In agriculture, autonomous tractors and drones equipped with sensors and AI algorithms are improving crop yields and optimizing resource usage. These real-world applications demonstrate the tangible benefits that autonomous robots can bring to different sectors and highlight the potential for increased efficiency, productivity, and innovation. While the integration of autonomous robots into various industries holds great promise, it also raises important ethical considerations that must be carefully addressed. Issues such as data privacy, cybersecurity, liability, and the impact on employment need to be carefully considered to ensure that the deployment of autonomous robots is beneficial for society as a whole. As autonomous robots become more advanced and autonomous, questions surrounding accountability and decision-making authority in case of accidents or errors must be clarified. The widespread adoption of autonomous robots can unlock myriad opportunities for enhancing human life and driving progress in the ever-evolving landscape of artificial intelligence.

XII. AI IN HEALTHCARE

The integration of artificial intelligence in healthcare has ushered in a new era of innovation and efficiency in the medical field. AI algorithms, combined with vast amounts of data, have the potential to revolutionize the way medical diagnoses are made and treatments are administered. By utilizing machine learning and deep learning techniques, AI systems can analyze complex medical data such as images, genetic information, and patient records to provide accurate, personalized healthcare solutions. This has the potential to improve patient outcomes, reduce medical errors, and optimize healthcare resources. One key area where AI is making a significant impact in healthcare is in medical imaging interpretation. AI-powered imaging software can analyze medical images such as X-rays, MRIs, and CT scans with incredible speed and accuracy, assisting radiologists in detecting abnormalities and making diagnoses. This not only speeds up the diagnostic process but also reduces the risk of misinterpretation and enhances the overall quality of patient care. AI algorithms can identify patterns and trends in patient data to predict potential health issues, allowing healthcare providers to intervene early and prevent serious complications. AI in healthcare is driving the development of autonomous medical robots that can assist in surgeries, deliver medications, and even provide remote patient monitoring. These robots can work collaboratively with healthcare professionals to streamline processes, increase efficiency, and improve overall patient care. While the integration of AI in healthcare presents exciting opportunities for improvement, it also raises ethical concerns regarding patient privacy, data security, and the potential for bias in algorithms. Addressing these challenges is crucial to ensure that the benefits of AI in healthcare are maximized while minimizing potential risks.

Role of AI in Medical Diagnosis

Due to significant advancements in artificial intelligence technology, the role of AI in medical diagnosis has experienced a remarkable transformation in recent years. AI algorithms are now being utilized to analyze vast amounts of medical data, including patient records, lab results, and imaging scans, to assist healthcare professionals in making accurate diagnoses. By leveraging machine learning techniques, AI systems can identify patterns and trends that may not be immediately apparent to human practitioners, thereby improving diagnostic accuracy and efficiency. This has the potential to revolutionize the field of medicine by enabling earlier detection of diseases and enhancing overall patient outcomes. AI-powered diagnostic tools have the ability to support healthcare providers in developing personalized treatment plans for patients based on their unique medical histories and genetic profiles. By integrating patient-specific data with evidence-based guidelines and clinical expertise, AI-driven diagnostic systems can offer tailored recommendations that optimize the effectiveness of medical interventions. This level of individualized care has the potential to significantly enhance patient satisfaction and outcomes, ultimately leading to a more efficient and effective healthcare system. The role of AI in medical diagnosis extends beyond the realm of individual patient care to population health management. By analyzing data from large patient cohorts, AI algorithms can identify

71

trends and risk factors associated with specific diseases, allowing for the implementation of targeted preventive measures and public health interventions. This proactive approach to healthcare can help reduce the burden of disease on healthcare systems and improve overall population health outcomes. As AI continues to evolve and become more sophisticated, its impact on medical diagnosis is poised to revolutionize healthcare delivery and drive advancements in precision medicine.

AI-Powered Medical Imaging Analysis

Emerging as a revolutionary application of artificial intelligence technologies, AI-powered medical imaging analysis has garnered significant attention in the healthcare industry. Leveraging machine learning algorithms, deep learning techniques, and advanced image processing, AI has the potential to transform the interpretation of medical images, such as X-rays, MRIs, and CT scans. By automating the analysis process, AI systems can assist radiologists in diagnosing diseases more accurately and efficiently, leading to improved patient outcomes. AI-powered medical imaging analysis can help in early detection of diseases, enabling prompt interventions and personalized treatment strategies. One of the key advantages of AI-powered medical imaging analysis is its ability to process vast amounts of imaging data quickly and accurately. Traditional manual interpretation of medical images can be time-consuming and prone to human errors. AI algorithms, on the other hand, can analyze thousands of images in a fraction of the time it would take a human expert, while maintaining a high level of accuracy. This speed and precision can be particularly crucial in emergency situations where swift diagnosis is essential for patient care. By augmenting the capabilities of healthcare professionals, AI-powered medical imaging analysis can streamline workflows, reduce diagnostic errors, and enhance overall operational efficiency within healthcare institutions. The implementation of AI-powered medical imaging analysis is not without challenges. Concerns about data privacy, security, regulatory compliance, and ethical considerations surrounding the use of AI in healthcare settings are paramount. Ensuring the transparency and interpretability of AI algorithms, addressing biases in training data, and integrating AI systems seamlessly into existing clinical workflows are key areas that require attention. Collaboration between healthcare providers, AI developers, regulatory bodies, and patients is essential to navigate these challenges and capitalize on the transformative potential of AI-powered medical imaging analysis in improving patient care and healthcare outcomes.

Ethical Considerations in AI Healthcare Applications

As artificial intelligence continues to revolutionize healthcare, ethical considerations become paramount in the development and implementation of AI applications in this field. One crucial ethical concern is the need to ensure patient privacy and data security. With the vast amount of sensitive information being processed by AI healthcare systems, there is a risk of data breaches and misuse if proper safeguards are not in place. It is imperative for developers and healthcare providers to prioritize the protection of patient data to maintain trust and compliance with privacy regulations. The issue of accountability and transparency in AI healthcare applications cannot be overlooked. As

AI systems make decisions that directly impact patient out-comes, it is crucial to understand how these decisions are made and hold those responsible for any errors or biases accountable. Transparent algorithms and clear guidelines for decision-mak-ing are essential to ensure that AI systems in healthcare are fair and just. The potential for AI to exacerbate existing healthcare disparities must be considered, and steps should be taken to mitigate biases and ensure equitable access to AI-driven healthcare services. The ethical implications of AI healthcare applications extend to the broader societal impact of these technologies. Questions surrounding the potential loss of human touch and empathy in healthcare delivery, as well as the ethical dilemmas posed by autonomous decision-making in critical sit-uations, need to be addressed. Balancing technological ad-vancements with ethical considerations is essential to ensure that AI in healthcare enhances patient care without compromis-ing key ethical principles. By addressing these ethical concerns proactively, stakeholders can foster public trust and maximize the benefits of AI in revolutionizing healthcare delivery.

XIII. AI IN FINANCE

As artificial intelligence continues to evolve, its application in the financial sector has become increasingly prominent. AI in finance is revolutionizing the way institutions manage risk, make investment decisions, and enhance customer service. Utilizing machine learning algorithms, AI systems can analyze vast amounts of data at a speed and accuracy that surpass human capabilities. This enables financial institutions to detect patterns, trends, and anomalies that may not be apparent to human analysts, leading to more informed decision-making processes. AI in finance plays a crucial role in fraud detection and prevention. By employing predictive analytics and pattern recognition, AI systems can flag suspicious activities in real-time, helping to mitigate potential risks and protect customer assets. AI-powered chatbots are transforming the customer service experience in the financial industry by providing personalized assistance and quick responses to inquiries. This not only improves customer satisfaction but also reduces operational costs for financial institutions. Looking ahead, the future of AI in finance holds immense promise, with the potential to streamline processes, enhance efficiency, and drive innovation. It is essential for financial institutions to address ethical considerations and regulatory challenges associated with the use of AI. As the technology continues to advance, striking a balance between harnessing its benefits and mitigating its risks will be crucial in ensuring the responsible and effective deployment of AI in the financial sector.

Applications of AI in Financial Services

In the realm of financial services, the applications of artificial intelligence have proven to be groundbreaking, revolutionizing the way institutions operate and how individuals interact with their finances. One prominent use case of AI in this sector is algorithmic trading, where machine learning algorithms analyze market trends and execute trades at speeds and frequencies impossible for human traders. This automated approach not only improves efficiency but also enhances decision-making by reducing human error and emotional biases. AI-powered chatbots are increasingly being utilized by banks and financial institutions to provide instant customer support, answer queries, and even offer personalized financial advice based on individual preferences and spending patterns. This not only enhances customer satisfaction but also streamlines processes and reduces operational costs for the organizations. AI is also being leveraged in the realm of fraud detection and prevention within the financial services industry. Machine learning algorithms can analyze vast amounts of data in real-time to identify suspicious patterns and flag potentially fraudulent activities, enabling institutions to take swift action to mitigate risks. By automating these processes and improving the accuracy of fraud detection, AI helps safeguard financial transactions and protect both institutions and their customers from financial losses. AI plays a crucial role in risk assessment and management, using predictive analytics to anticipate potential risks and optimize investment strategies. This proactive approach enables financial institutions to make more informed decisions, allocate resources effectively, and mitigate risks before they escalate, ultimately enhancing stability and resilience in the ever-changing financial landscape.

The applications of AI in financial services have transformed the industry, driving innovation, efficiency, and customer-centric approaches. From algorithmic trading to personalized customer service and fraud detection, AI has revolutionized traditional practices, enabling institutions to navigate complex challenges and seize new opportunities. As technology continues to advance, the role of AI in financial services is expected to grow, reshaping business models, enhancing decision-making capabilities, and ultimately redefining the relationship between financial institutions and their clients. By embracing and harnessing the power of AI, the financial services sector stands to benefit from increased operational efficiency, improved risk management, and a more personalized and streamlined customer experience.

Algorithmic Trading and Risk Management

One critical aspect of artificial intelligence that has gained increasing attention in recent years is algorithmic trading and risk management. Algorithmic trading refers to the use of computer algorithms to execute trading decisions at high speeds, leveraging vast amounts of data and sophisticated mathematical models. This approach has revolutionized financial markets by enabling rapid transactions, minimizing human errors, and capturing opportunities that may be missed by manual trading. With the automation of trading processes comes inherent risks, such as market volatility, algorithmic errors, and potential system failures. As a result, robust risk management strategies are essential to mitigate these risks and ensure the stability and integrity of financial markets. Risk management in algorithmic trading encompasses a range of techniques and tools designed to identify, assess, and address potential risks associated with automated trading systems. These include pre-trade risk controls to limit the size and frequency of trades, post-trade analysis to evaluate the impact of trading decisions, and stress testing to simulate extreme market conditions. Risk management practices in algorithmic trading often involve monitoring and surveillance mechanisms to detect anomalies or irregularities in trading behavior, as well as compliance with regulatory requirements to ensure transparency and accountability. By integrating these risk management measures into algorithmic trading systems, financial institutions can enhance their operational resilience and protect against potential market disruptions. While algorithmic trading has revolutionized financial markets by increasing efficiency and liquidity, it also poses unique challenges and risks that must be carefully managed. Effective risk management practices in algorithmic trading require a comprehensive understanding of market dynamics, technological capabilities, and regulatory frameworks. By adopting a proactive approach to risk management, financial institutions can maintain market integrity, safeguard investor interests, and foster a more stable and sustainable trading environment. As AI continues to advance and evolve, the importance of robust risk management in algorithmic trading will only become more pronounced, underscoring the need for ongoing vigilance and adaptation in the face of rapid technological change.

Regulatory Challenges in AI-driven Finance

One of the significant challenges in the realm of AI-driven finance is the lack of clear regulatory frameworks to govern the use of artificial intelligence algorithms in financial decision-making processes. As AI algorithms become more sophisticated and complex, the risks associated with their use, such as potential biases and data privacy concerns, become more pronounced. Without proper regulations in place, there is a risk of algorithmic decision-making leading to unintended consequences and financial instability. Regulatory bodies face the challenge of keeping pace with the rapid advancements in AI technology while also ensuring that financial institutions uphold transparency, fairness, and accountability in their use of AI. The cross-border nature of AI-driven financial systems adds another layer of complexity to regulatory challenges. With data being transmitted and processed across different jurisdictions, regulators must grapple with issues related to data localization, data sovereignty, and harmonization of regulations. The lack of international standards for AI in finance exacerbates the regulatory hurdles faced by policymakers and can lead to regulatory arbitrage and inconsistencies in compliance requirements. Harmonizing regulations across borders to address the unique challenges posed by AI-driven finance will require collaboration and coordination among regulatory bodies at the international level. The rapid evolution of AI technologies in finance presents challenges in monitoring and enforcing compliance with existing regulations. Traditional regulatory frameworks may not fully capture the nuanced risks associated with AI-driven financial products and services, such as algorithmic biases and explainability of AI decisions. Regulators must adapt and update existing regulations to address the specific risks posed by AI in finance, such as the need for algorithmic transparency, model validation, and ongoing monitoring of AI systems. Developing regulatory sandboxes and frameworks tailored to AI technologies can help regulators stay abreast of emerging risks and foster innovation in the financial sector while ensuring consumer protection and financial stability.

XIV. AI IN EDUCATION

The integration of artificial intelligence into the field of education has the potential to revolutionize traditional teaching methods and enhance the learning experience for students. By utilizing AI-powered tools such as personalized learning platforms, virtual tutors, and intelligent grading systems, educators can effectively tailor instruction to meet the individual needs and preferences of each student. This personalized approach can lead to improved student engagement, retention, and academic performance. AI can analyze vast amounts of student data to identify patterns and trends, allowing educators to make data-driven decisions to effectively support student learning and development. AI in education has the power to democratize access to quality education by breaking down barriers such as geographical location, socioeconomic status, or physical disabilities. Virtual classrooms powered by AI technology can provide students with remote access to high-quality educational resources, expert teachers, and interactive learning experiences. This flexibility in learning modalities can accommodate diverse learning styles and preferences, catering to a wider range of students who may not thrive in traditional classroom settings. By leveraging AI in education, institutions can foster a more inclusive and equitable learning environment for all students, ensuring that everyone has the opportunity to reach their full academic potential. While the potential benefits of AI in education are vast, there are also ethical considerations and challenges that must be carefully considered. Issues such as data privacy and security, bias in algorithms, and the displacement of teachers by AI-powered systems need to be addressed to ensure that the integration of AI in education is conducted ethically and responsibly. There is a need for ongoing professional development for educators to build their capacity to effectively integrate AI tools into their teaching practices. By navigating these complex challenges thoughtfully, the field of education can harness the power of AI to create innovative and impactful learning experiences that benefit students, teachers, and society as a whole.

Personalized Learning with AI

The evolution of artificial intelligence has paved the way for personalized learning with the integration of AI technology. By leveraging AI algorithms, personalized learning platforms can provide tailored educational experiences for students based on their individual needs, preferences, and learning styles. These platforms can adapt to the pace of each student, offering targeted content and resources to enhance their understanding and engagement. AI-powered systems can analyze vast amounts of data to identify trends and patterns in student performance, allowing for more effective interventions and support mechanisms to be put in place. Personalized learning with AI enables educators to shift their focus from traditional, one-size-fits-all instruction to a more student-centered approach. By harnessing AI capabilities, teachers can access real-time data on student progress, enabling them to track individual growth, identify areas of weakness, and tailor their teaching strategies accordingly. This data-driven approach not only enhances the effectiveness of instruction but also empowers educators to provide personalized feedback and support to each student, fostering a more inclusive and responsive learning environment. The integration of AI in personalized learning opens up new possibilities for the future of education. As AI technology continues to advance, students can benefit from adaptive learning systems that continuously evolve and improve based on individual feedback and performance. This dynamic and interactive approach to education has the potential to revolutionize traditional learning models, promoting greater student engagement, autonomy, and success. By embracing personalized learning with AI, educational institutions can empower students to take ownership of their learning journey and ultimately unlock their full potential.

AI Tutoring Systems

Advances in artificial intelligence have paved the way for the development of AI tutoring systems, which are revolutionizing the way individuals learn and acquire new skills. These systems leverage AI technologies such as machine learning and natural language processing to provide personalized and adaptive learning experiences. By analyzing user interactions and performance data, AI tutoring systems can tailor their instructional content to meet the unique needs and learning styles of each student. This level of customization ultimately leads to more effective learning outcomes and greater engagement among learners. AI tutoring systems have the ability to provide immediate feedback and support to students, helping them address their learning gaps in real time. This instant feedback mechanism not only enhances the learning experience but also promotes a deeper understanding of the material being taught. AI tutoring systems can track the progress of each student and adjust their instructional approach accordingly, ensuring that learners stay on track and achieve their educational goals. This personalized approach to learning is reshaping traditional education models and opening up new opportunities for individuals to access high-quality tutoring services at scale. As AI tutoring systems continue to evolve and improve, they hold the potential to democratize education by making high-quality learning resources more accessible and affordable to a wider audience. By leveraging the power of AI, these systems can provide personalized tutoring services to students regardless of their location or socio-economic background. This democratization of education has the potential to narrow the achievement gap and empower individuals to reach their full potential. AI tutoring systems represent a significant advancement in the field of education, offering a glimpse into the future of personalized and adaptive learning experiences.

Ethical Implications of AI in Education

As artificial intelligence continues to infiltrate various aspects of society, its integration in education comes with a multitude of ethical implications that must be carefully considered. One key concern revolves around the potential for AI to exacerbate existing inequalities in education. The implementation of AI-powered tools in classrooms may inadvertently disadvantage students who do not have access to the necessary technology or lack the digital literacy skills needed to navigate these systems effectively. This could widen the gap between students from different socioeconomic backgrounds, further perpetuating disparities in educational outcomes. Another ethical consideration in the realm of AI in education involves privacy and data security. With the use of AI platforms to collect and analyze vast amounts of student data, there is a legitimate concern regarding the protection of sensitive information. The risk of breaches or misuse of personal data raises questions about who has access to this information, how it is being used, and whether students and parents have adequate control over their own data. As educational institutions increasingly rely on AI algorithms to personalize learning experiences, ensuring the ethical handling of data becomes paramount to safeguarding the privacy and rights of students. The potential for AI in education to influence decision-making processes raises ethical dilemmas surrounding accountability and transparency. As algorithms play a more significant role in assessing student performance, recommending courses, or even identifying potential career paths, questions arise about the fairness and bias inherent in these systems. Issues of algorithmic transparency, explainability, and the ability to challenge or appeal automated decisions become crucial in maintaining accountability and ensuring that the educational opportunities provided through AI are equitable for all students. By addressing these ethical implications proactively, stakeholders can harness the power of AI in education while upholding principles of fairness, privacy, and social responsibility.

XV. AI IN ENVIRONMENTAL SUSTAINABILITY

As technology continues to advance, the role of artificial intelligence in environmental sustainability is becoming increasingly significant. AI has the potential to revolutionize how we address environmental challenges by enhancing efficiency, improving resource management, and optimizing decision-making processes. By utilizing AI-powered tools and technologies, organizations and governments can better monitor environmental changes, predict natural disasters, and develop innovative solutions to mitigate the impact of climate change. This proactive approach enabled by AI can help create a more sustainable future for our planet. One key area where AI can make a substantial impact on environmental sustainability is in the field of renewable energy. By leveraging AI algorithms to analyze vast amounts of data, researchers and engineers can optimize the design and operation of renewable energy systems such as solar panels, wind turbines, and hydroelectric plants. AI can also be used to forecast energy demand, optimize energy consumption patterns, and facilitate the integration of renewable energy sources into existing power grids. This not only reduces greenhouse gas emissions but also promotes the transition towards a more sustainable and resilient energy infrastructure. AI can play a crucial role in promoting sustainable practices in industries such as agriculture, transportation, and waste management. From precision agriculture techniques that optimize crop yields while minimizing environmental impact to autonomous vehicles that reduce fuel consumption and traffic congestion, AI-powered applications are revolutionizing traditional practices. By harnessing the power of AI to collect, analyze, and act on environmental data, stakeholders can make informed decisions that lead to more sustainable outcomes. As we continue to harness the potential of AI in environmental sustainability, it is essential to prioritize collaboration, transparency, and ethical considerations to ensure that these technologies are used responsibly for the benefit of both present and future generations.

AI for Climate Change Prediction

Advancements in artificial intelligence have opened up new horizons in the realm of climate change prediction. AI technologies, particularly machine learning algorithms, are being leveraged to analyze vast amounts of environmental data and predict future climate trends with a high degree of accuracy. By processing data from various sources such as satellites, weather stations, and ocean buoys, AI systems can identify patterns and correlations that human analysts may overlook. This has the potential to revolutionize our understanding of climate change and enable policymakers to make informed decisions to mitigate its impact. One of the key advantages of using AI for climate change prediction is its ability to handle large and complex datasets in real-time. Traditional methods of climate modeling often struggle to process the sheer volume of data available, leading to limitations in accuracy and timeliness. AI algorithms, on the other hand, excel at handling such data-intensive tasks, allowing researchers to generate more precise and timely predictions. This capability is crucial in addressing the urgent need for accurate climate change forecasts to guide mitigation efforts and adaptation strategies at both global and local levels. AI can also help identify and assess the effectiveness of various climate change interventions, such as renewable energy projects, carbon sequestration initiatives, and conservation efforts. By analyzing past data and running simulations based on different scenarios, AI systems can predict the potential outcomes of these interventions and provide valuable insights to policymakers and stakeholders. This proactive approach to climate change management can help prioritize resources and optimize strategies for maximum impact. As AI continues to evolve, its role in climate change prediction and mitigation is poised to become even more indispensable in shaping a sustainable future for our planet.

Smart Energy Management Systems

As smart energy management systems become more prevalent in our daily lives, the integration of artificial intelligence has played a crucial role in optimizing energy consumption and increasing efficiency. These systems leverage AI algorithms to analyze data from various sources, such as smart meters and IoT devices, to make intelligent decisions in real-time. By continuously monitoring energy usage patterns and detecting anomalies, AI-powered systems can automatically adjust settings to minimize waste and reduce costs. This proactive approach not only ensures a more sustainable energy consumption but also contributes to a more environmentally friendly and cost-effective energy management strategy. One key benefit of smart energy management systems enhanced by AI is their ability to predict future energy demands and trends. By analyzing historical data and patterns, these systems can forecast energy consumption needs and optimize energy production and distribution accordingly. This predictive capability allows for better planning and resource allocation, ultimately leading to a more stable and reliable energy supply. AI algorithms can identify opportunities for energy conservation and efficiency improvements, helping organizations and individuals make informed decisions to reduce their carbon footprint and save on energy costs. The evolution of smart energy management systems with AI has also opened up new possibilities for demand response and grid optimization. By leveraging real-time data and advanced analytics, these systems can adjust energy consumption based on external factors such as weather conditions, market prices, and grid constraints. This dynamic approach not only helps to balance supply and demand more effectively but also enables users to participate in demand-side management programs and optimize their energy consumption patterns in alignment with grid requirements. The integration of AI into smart energy management systems holds great promise for achieving a more sustainable and efficient energy ecosystem for the future.

Ethical Considerations in AI for Environmental Conservation

Ethical considerations in the development and implementation of AI for environmental conservation are crucial as this technological advancement continues to reshape our approach to addressing environmental challenges. One key consideration revolves around the potential impact of AI on employment, particularly in sectors traditionally reliant on manual labor for conservation efforts. As AI becomes more integrated into conservation practices, there is a risk of displacing human workers, raising questions about equitable access to opportunities in the emerging AI-driven conservation industry. Ensuring a just transition for workers affected by automation and investing in retraining programs will be essential to mitigate the negative social implications of AI adoption in environmental conservation. Ethical concerns also extend to the data privacy and security aspects of AI applications in conservation. With the vast amounts of data collected and analyzed by AI systems to monitor and manage natural resources, there is a heightened risk of data breaches and unauthorized access to sensitive environmental information. Striking a balance between leveraging data-driven insights for effective conservation outcomes and safeguarding the privacy and security of individuals and ecosystems will be a paramount ethical consideration in the advancement of AI for environmental conservation. Robust data governance frameworks and adherence to ethical standards in data management practices will be essential to build trust and ensure the responsible use of AI technologies in conservation efforts. Transparency and accountability in the development and deployment of AI systems for environmental conservation are essential to maintain public trust and uphold ethical standards. As AI algorithms increasingly influence decision-making processes in conservation, there is a need for clear guidelines and mechanisms to ensure that these systems operate in a fair and unbiased manner. Addressing issues such as algorithmic bias and ensuring the explainability of AI-driven conservation decisions will be critical to promoting ethical AI practices that align with principles of fairness, accountability, and transparency.

XVI. AI IN CREATIVE INDUSTRIES

As technology continues to advance, artificial intelligence has made significant inroads into various industries, including the creative sector. AI in creative industries is revolutionizing the way artists, designers, and content creators work, offering new tools and possibilities for innovation. From generating new ideas to assisting in the production process, AI is playing a crucial role in reshaping the creative landscape. One key aspect of AI in creative industries is its ability to analyze vast amounts of data and identify patterns that humans may overlook. This can be particularly useful in fields such as marketing and advertising, where understanding consumer behavior and preferences is essential. By leveraging AI algorithms, businesses can gain deeper insights into their target audience and tailor their creative strategies accordingly. This not only enhances the effectiveness of marketing campaigns but also enables companies to stay ahead of trends and competitors. AI-powered tools are enabling artists and designers to experiment with new techniques and push the boundaries of creativity. AI algorithms can assist in generating unique visual effects, music compositions, or design concepts that artists can then refine and expand upon. By incorporating AI into their creative process, professionals in the creative industries can explore new possibilities, collaborate with intelligent systems, and ultimately deliver more innovative and engaging content to audiences. As AI continues to evolve, its impact on creative industries will only grow, opening up new avenues for expression and artistic exploration.

AI-generated Art and Music

The emergence of artificial intelligence has not only revolutionized the way we interact with technology but has also expanded
into the realms of art and music. AI-generated art has gained
traction in recent years, with algorithms being used to create
unique and intricate pieces across different mediums. From
paintings to sculptures, AI has showcased its ability to mimic
human creativity and produce works that challenge traditional
notions of artistry. Similarly, AI-generated music has also made
waves in the music industry, with algorithms composing melo-
dies that evoke emotion and captivate listeners. The marriage
of AI and art has opened up new possibilities for creativity and
artistic expression, blurring the lines between man and machine.
One of the key advantages of AI-generated art and music lies
in its ability to push the boundaries of creativity and explore
uncharted territories. By harnessing the power of machine learning and neural networks, AI can analyze vast amounts of data
to produce art and music that goes beyond human imagination.
This innovative approach not only challenges conventional artistic practices but also introduces new avenues for experimentation and innovation. AI-generated art and music have the potential to inspire artists and musicians to think outside the box
and explore novel ways of expressing themselves, ultimately en-
riching the cultural landscape with fresh perspectives and innovative creations. Despite the groundbreaking advancements in
AI-generated art and music, there are concerns surrounding the
authenticity and originality of these works. Critics argue that AI
lacks the inherent emotions and experiences that drive human
creativity, raising questions about the true artistic value of AI-
generated pieces. There are ethical considerations regarding the

role of AI in the creative process and its impact on human artists and musicians. As the line between man and machine continues to blur, it is essential to engage in thoughtful discussions about the implications of AI-generated art and music on the future of creativity and artistic expression. By balancing technological innovation with ethical considerations, we can navigate the evolving landscape of AI-generated art and music with a critical eye towards preserving the integrity and authenticity of human creativity.

AI in Content Creation and Curation

Advances in artificial intelligence have significantly impacted the realm of content creation and curation. With the development of sophisticated algorithms and machine learning techniques, AI has transformed the way content is generated, curated, and shared. Virtual assistants, such as Siri, Alexa, and Google Assistant, represent an initial step towards AI-powered content creation. These virtual assistants have become integral parts of modern life, providing users with personalized information, recommendations, and assistance. Their ability to understand natural language and respond to complex queries has revolutionized the way people interact with technology, leading to increased efficiency and convenience in various industries. As AI continues to advance, the scope of content creation and curation expands to include autonomous robots. These robots are capable of performing tasks without human intervention, ranging from autonomous vehicles to cleaning robots and industrial automation. The development of autonomous robots raises both technological and ethical challenges, as they have the potential to revolutionize the way businesses operate and society functions as a whole. The use of autonomous robots in sectors such as medicine, logistics, agriculture, and security showcases the diverse applications of AI in content creation and curation, further emphasizing the growing importance of these technologies in shaping the future. Looking forward, the evolution of AI from virtual assistants to autonomous robots points towards a future where artificial intelligence will play an even more dominant role in content creation and curation. Emerging trends in AI and robotics suggest a continued integration of these technologies into various aspects of human life, with potential benefits and risks to consider. As society prepares for a future driven by AI, it is crucial to strike a balance between technological innovation and ethical considerations, ensuring that the widespread adoption of autonomous robots and AI-powered content creation serves the greater good.

Intellectual Property Rights and AI-generated Content

As artificial intelligence continues to evolve, one of the key areas of concern is the question of intellectual property rights in relation to AI-generated content. With the increasing capabilities of AI to create music, art, literature, and other forms of creative work, the issue of who owns the rights to these creations becomes complex. Traditional copyright laws are designed to protect the works of human creators, but they may not be well-equipped to address the unique challenges posed by AI-generated content. The crux of the issue lies in determining whether AI should be considered the creator of the content it produces, or if the individuals or organizations that develop and deploy the AI should be considered the rightful owners of the works generated. Some argue that since AI operates based on algorithms and data inputs provided by humans, the ultimate creative contribution still comes from human beings. They posit that the intellectual property rights should reside with the creators of the AI system rather than the AI itself. This perspective raises questions about accountability and attribution in cases where AI autonomously generates content without direct human intervention. In order to address the complexity of intellectual property rights in the context of AI-generated content, policymakers, legal scholars, and industry stakeholders need to engage in thorough discussions and develop clear guidelines. It is essential to strike a balance that recognizes and rewards human creativity while also allowing for the potential of AI to generate innovative and valuable content. As AI technology continues to advance, the need for establishing robust frameworks that protect the rights of creators, whether human or artificial, will become increasingly urgent to ensure fair and equitable outcomes for all parties involved.

XVII. AI GOVERNANCE AND ACCOUNTABILITY

As artificial intelligence continues to advance rapidly, the issue of governance and accountability becomes increasingly crucial. With the development of more sophisticated AI systems, there is a growing need to establish clear guidelines and regulations to ensure that these technologies are used ethically and responsibly. AI governance involves setting standards for the design, development, and deployment of AI systems, while accountability ensures that those responsible for AI-related decisions can be held answerable for their actions. Without proper governance and accountability measures in place, there is a risk that AI systems could be misused or cause harm to individuals or society as a whole. One of the key challenges in AI governance is the lack of existing regulatory frameworks that are equipped to address the complexities of AI technologies. Traditional laws and regulations may not be sufficient to govern AI systems that learn and adapt over time, posing unique ethical and legal dilemmas. The decentralized nature of AI development, with tech companies and research institutions around the world contributing to advancements, further complicates the task of establishing global standards for AI governance. As such, there is a pressing need for international collaboration and cooperation to develop a framework that can effectively regulate the use of AI while promoting innovation and development. In order to ensure accountability in AI decision-making processes, transparency and explainability of AI systems are essential. This means that stakeholders should have a clear understanding of how AI algorithms make decisions and the potential biases or risks associated with these decisions. Implementing mechanisms for auditing and monitoring AI systems can help detect and correct errors or biases, holding developers and users accountable for the outcomes of AI applications. By fostering a culture of accountability in AI development and deployment, we can promote trust and confidence in these technologies while safeguarding against potential misuse or harm.

Importance of AI Governance Frameworks

As artificial intelligence continues to evolve and permeate various aspects of our lives, the importance of implementing effective governance frameworks becomes increasingly crucial. These frameworks serve as a set of guidelines and principles that regulate the development, deployment, and use of AI technologies to ensure ethical, safe, and responsible practices. Without proper governance, the potential risks associated with AI, such as bias, privacy concerns, and security threats, can overshadow its benefits. As AI systems become more sophisticated and autonomous, the need for robust governance mechanisms becomes even more pressing to mitigate potential harms and ensure that AI aligns with societal values and objectives. One key aspect of AI governance frameworks is the establishment of accountability mechanisms to hold developers and deployers of AI technologies responsible for their actions. By clearly defining roles and responsibilities, setting standards for transparency and explainability, and implementing mechanisms for oversight and accountability, governance frameworks can help prevent misuse or abuse of AI systems. These frameworks can facilitate compliance with legal and regulatory requirements, ensuring that AI applications adhere to relevant laws and guidelines. This accountability helps build trust among stakeholders, including users, policymakers, and the public, fostering a positive perception of AI technologies and encouraging their responsible use and adoption. AI governance frameworks play a vital role in promoting fairness, equity, and inclusion in the development and deployment of AI technologies. By addressing issues of bias, discrimination, and social impact, these frameworks can help mitigate the negative consequences of AI systems on individuals and communities. Through mechanisms such as fairness audits, impact assessments, and stakeholder engagement, governance frameworks can help identify and address potential biases and discrimination in AI algorithms and applications. By promoting diversity, equity, and inclusion in AI development teams and decision-making processes, these frameworks can help ensure that AI technologies serve the needs and interests of a diverse range of stakeholders, ultimately contributing to more equitable and inclusive outcomes.

Transparency and Accountability in AI Systems

One of the key challenges in the field of artificial intelligence systems is ensuring transparency and accountability in their operations. As AI systems become more complex and autonomous, it becomes increasingly important to understand how they make decisions and why. Transparency refers to the ability to trace and understand the decision-making process of AI systems, while accountability is the concept of holding these systems responsible for their actions. Without transparency, it is difficult to trust AI systems, especially when they are used in critical applications such as healthcare or transportation. Accountability, on the other hand, ensures that AI systems can be held responsible for any errors or biases in their decision-making, thus fostering trust and reliability in their use. In order to achieve transparency and accountability in AI systems, it is essential to implement mechanisms that enable users to understand how these systems arrive at their decisions. This can be achieved through explainable AI, which focuses on developing models and algorithms that provide clear explanations for their outputs. By enabling humans to understand the reasoning behind AI decisions, trust and confidence in these systems can be established. Accountability can be ensured through proper governance frameworks and regulations that hold developers and users of AI systems accountable for their actions. By establishing clear guidelines and standards for the ethical use of AI, the potential risks and harmful outcomes of these systems can be minimized. Transparency and accountability are essential components of ethical AI development and deployment. By prioritizing transparency, developers can enhance the interpretability of AI systems, making them more trustworthy and reliable. Simultaneously, accountability mechanisms can help mitigate the risks associated with AI systems, ensuring that they are used responsibly and ethically. As AI continues to advance and integrate into various aspects of society, ensuring transparency and accountability will be crucial in harnessing its benefits while minimizing its potential harms.

International Collaboration for AI Regulation

International collaboration is crucial for the effective regulation of artificial intelligence technologies. Given the global nature of

AI development and deployment, a unified approach to regulation is essential to ensure consistency and coherence across borders. Countries must work together to establish common standards and guidelines that can govern the ethical use of AI, protect user privacy, and mitigate potential risks associated with these technologies. By fostering international collaboration, policymakers can create a more predictable regulatory environment that fosters innovation while safeguarding societal well-being. One key benefit of international collaboration in AI regulation is the ability to share best practices and lessons learned. Different countries may have varying levels of expertise and experience in regulating AI technologies, and by working together, they can leverage each other's knowledge to develop more effective regulatory frameworks. This can help avoid duplication of efforts and ensure that regulations are based on evidence and experience rather than trial and error. International collaboration can enable countries to address common challenges in a coordinated manner, such as preventing AI-driven discrimination or ensuring transparency in AI decision-making processes. International collaboration can help build trust and credibility in AI technologies by demonstrating a commitment to ethical principles and societal values. By aligning on common standards for AI regulation, countries can send a strong signal to the public and industry that they take the responsible development and deployment of AI seriously. This can help foster greater trust in AI technologies and encourage their acceptance and adoption by users and businesses. By working together on AI regulation, countries can build a more robust and resilient regulatory framework that promotes innovation while protecting individuals and society as a whole.

XVIII. AI AND HUMAN AUGMENTATION

Advancements in artificial intelligence have led to the integration of human augmentation technologies that aim to enhance human capabilities and performance. The synergy between AI and human augmentation has the potential to revolutionize various industries and fields, creating a new paradigm where humans and machines work together to achieve optimal results. By leveraging AI technologies such as machine learning and neural networks, human augmentation tools can analyze vast amounts of data to provide personalized insights and recommendations, enabling individuals to make informed decisions and improve their skills and abilities. One prominent example of the fusion between AI and human augmentation is the development of exoskeletons that can amplify human strength and endurance. These wearable robotic devices utilize AI algorithms to interpret the user's movements and provide the necessary support to enhance physical performance. By leveraging AI-powered sensors and actuators, exoskeletons can adapt in real-time to the user's needs, offering assistance in tasks that require precision and control. This seamless integration between human users and intelligent machines demonstrates the potential for AI to augment human capabilities and redefine the boundaries of what is possible. The convergence of AI and human augmentation also raises ethical considerations regarding the implications of enhancing human abilities through technological means. As these technologies become more widespread and accessible, questions arise regarding equity, privacy, and consent in the use of AI-powered human augmentation tools. It is essential to establish robust ethical frameworks and regulations to ensure that the integration of AI and human augmentation is undertaken responsibly and ethically, prioritizing the well-being and autonomy of individuals. By addressing these ethical challenges proactively, society can harness the full potential of AI and human augmentation to improve human performance and quality of life while upholding fundamental values and principles.

Integration of AI with Human Abilities

As technology continues to advance, the integration of artificial intelligence with human abilities has become a prominent topic of discussion. This fusion of AI and human capabilities has the potential to revolutionize various industries and aspects of everyday life. One of the key aspects of this integration is the development of virtual assistants, such as Siri, Alexa, and Google Assistant, which have become indispensable tools in helping individuals perform tasks more efficiently. These virtual assistants have evolved from simple speech recognition tools to sophisticated programs capable of understanding and responding to complex commands, blurring the lines between human and machine interaction. The progression from virtual assistants to autonomous robots marks the next frontier in AI technology. Autonomous robots, equipped with the ability to operate independently and make decisions based on their programming and environment, are increasingly being utilized in sectors such as transportation, manufacturing, and healthcare. These robots have the potential to streamline processes, increase efficiency, and reduce the risk of human error in various industries. The development of autonomous robots also raises ethical and technological challenges that must be addressed to ensure their safe and responsible integration into society. The integration of AI with human abilities is not without its complexities and implications. As autonomous robots become more prevalent in our daily lives, it is essential to consider the potential benefits and risks associated with their widespread adoption. From improving business efficiency to transforming healthcare delivery, the impact of AI integration will continue to reshape how tasks are performed and decisions are made. As we navigate this evolving landscape, it is crucial to strike a balance between technological innovation and ethical considerations to ensure that AI enhances human capabilities while safeguarding privacy, security, and societal values.

Ethical Implications of Human-AI Collaboration

As human-AI collaboration becomes more prevalent in various fields, there are several ethical implications that need to be carefully considered. One major concern is the issue of accountability and responsibility. When AI systems make decisions or carry out tasks in conjunction with humans, it can become challenging to determine who is ultimately responsible when something goes wrong. This blurring of lines between human agency and AI autonomy raises questions about liability and the allocation of accountability in cases of error or harm caused by AI. Establishing clear guidelines and frameworks for attributing responsibility in human-AI collaborations is crucial to ensure ethical decision-making and address potential legal issues that may arise. Human-AI collaboration also raises concerns about privacy and data security. As AI systems collect and analyze vast amounts of data to facilitate collaboration with humans, there is a risk of privacy breaches and unauthorized access to sensitive information. Ensuring the protection of personal data and upholding individual privacy rights are essential considerations in the development and deployment of AI technologies. Striking a balance between the need for data-driven insights and the ethical principles of privacy and confidentiality is key to maintaining trust and integrity in human-AI interactions. Implementing robust data protection measures and transparent data governance practices can help mitigate these risks and safeguard individuals' rights in the era of AI collaboration. There is a need to address issues of bias and discrimination in human-AI collaboration. AI systems are only as unbiased and fair as the data they are trained on, which can reflect and perpetuate existing societal biases and prejudices. When humans work alongside AI technologies, there is a risk of amplifying and institutionalizing these biases in decision-making processes. It is crucial to critically examine and mitigate bias in AI algorithms and ensure that human-AI collaborations promote fairness, equity, and inclusivity. Implementing bias detection mechanisms, diversifying datasets, and fostering diverse perspectives in AI development are essential steps to advance ethical considerations in human-AI collaboration and prevent discriminatory outcomes.

Future Prospects of Human-AI Integration

The potential for human-AI integration in the future is both promising and complex. As virtual assistants continue to evolve, moving from basic functionalities to more advanced natural language understanding, individuals are becoming increasingly reliant on AI systems for everyday tasks. This trend is likely to continue as technology progresses, paving the way for more seamless interactions between humans and machines. The integration of machine learning and neural networks has significantly enhanced the capabilities of virtual assistants, enabling them to comprehend and respond to complex commands with greater accuracy, further bridging the gap between human communication and machine understanding. Looking ahead, autonomous robots represent the next frontier in artificial intelligence, offering a glimpse into a future where intelligent machines can operate independently in various contexts. From autonomous vehicles revolutionizing transportation to cleaning robots enhancing household chores, the potential applications of autonomous robots are vast. The development of autonomous robots is not without challenges, with technological and ethical considerations coming to the forefront. As society embraces these advanced technologies, questions surrounding safety, accountability, and job displacement will need to be addressed to ensure a responsible and ethical integration of autonomous robots into our daily lives. Despite the inevitable uncertainties and challenges that come with the increasing integration of AI into human society, the future of artificial intelligence holds immense promise for various industries and sectors. As autonomous robots become more prevalent, their impact on business efficiency, productivity, and innovation will be profound. The potential benefits of widespread AI adoption are significant, but careful consideration of the risks and ethical implications is crucial. In navigating this evolving landscape, it is imperative for policymakers, industry leaders, and the public to collaborate in shaping a future that maximizes the benefits of AI integration while safeguarding against potential pitfalls.

XIX. AI AND GLOBAL SECURITY

The development of artificial intelligence has raised significant concerns regarding global security. As AI systems become more advanced and autonomous, there is a growing fear of the potential misuse of this technology for malicious purposes. One of the primary concerns is the use of AI in cyber warfare, where sophisticated AI-powered attacks could disrupt critical infrastructure, financial systems, and government operations. The ability of AI to quickly analyze vast amounts of data and adapt its tactics makes it a powerful tool for both offensive and defensive cyber operations, posing a significant challenge for cybersecurity professionals worldwide. The deployment of autonomous robots in military operations has also sparked debates surrounding the ethical implications of using machines to make life-and-death decisions. The development of autonomous weapons systems, capable of identifying and engaging targets without human intervention, raises concerns about the potential for accidental killings, escalation of conflicts, and the erosion of human control over warfare. The lack of clear regulations and guidelines regarding the use of AI in weapons systems further complicates the issue, prompting calls for international treaties to ban or restrict the use of autonomous weapons to prevent destabilizing arms races and indiscriminate violence. In response to these challenges, there is a growing recognition of the need for collaborative efforts between governments, international organizations, and technology companies to establish frameworks for the responsible development and deployment of AI technologies. Ethical guidelines, transparency measures, and accountability mechanisms are essential to ensure that AI is used in a manner that upholds fundamental human rights, promotes peace, and enhances global security. By fostering dialogue and cooperation on AI governance, stakeholders can work together to harness the potential of AI for positive advancements while mitigating its risks and safeguarding the future of humanity.

AI in Cybersecurity

Artificial intelligence has become an essential tool in the field of cybersecurity, revolutionizing the way organizations protect their digital assets. By leveraging AI technologies, security professionals can detect and respond to threats in real-time, allowing for a proactive approach to cybersecurity rather than a reactive one. One of the key benefits of AI in cybersecurity is its ability to analyze vast amounts of data quickly and accurately, identifying potential threats that may go unnoticed by traditional security measures. This not only enhances the overall security posture of an organization but also helps in mitigating risks and preventing potential breaches before they occur. AI-driven cybersecurity solutions can adapt and evolve based on new threats and trends, constantly learning and improving their detection capabilities. This level of adaptability is crucial in today's ever-evolving threat landscape, where cyberattacks are becoming more sophisticated and frequent. AI can automate routine security tasks, freeing up human analysts to focus on more strategic initiatives and investigations. By streamlining operations and enhancing efficiency, organizations can better defend against cyber threats while optimizing their resources effectively. The integration of AI in cybersecurity is paving the way for the development of autonomous defense systems that can proactively identify, respond to, and neutralize threats without human intervention. This level of autonomy not only improves the speed and accuracy of threat detection but also reduces the response time to incidents, minimizing the potential impact of cyberattacks. As organizations continue to rely on AI for cybersecurity, it is crucial to address ethical considerations and ensure that such technologies are deployed responsibly to maintain transparency, accountability, and integrity in the protection of sensitive data and information.

Autonomous Weapons Systems

As autonomous weapons systems continue to gain attention in the realm of artificial intelligence, ethical concerns around their use and development have surfaced. The concept of autonomous weapons systems refers to weapons that can operate without direct human intervention, making decisions independently based on pre-programmed algorithms. This raises questions about accountability and potential loss of human control over lethal force, which could have significant consequences in conflict situations. The rapid advancement of technology in this field also brings about the need for international regulations to ensure the responsible deployment and use of autonomous weapons systems. One of the key challenges associated with autonomous weapons systems is the ethical dilemma of assigning decision-making power to machines in life-and-death scenarios. The ability of these systems to react faster than humans and potentially engage in combat without human oversight raises concerns about unintended consequences and the potential for escalation. The lack of emotional intelligence and moral reasoning capabilities in machines further complicates the ethical considerations of allowing autonomous weapons systems to make critical decisions in complex and dynamic environments. As such, the development of robust ethical frameworks and guidelines is crucial to mitigate the risks associated with autonomous weapons systems. The proliferation of autonomous weapons systems has implications beyond the battlefield, extending to the realm of national security and global stability. The potential for the use of these systems in asymmetric warfare, terrorism, and other non-state actor activities raises the stakes for policymakers and international actors. The need for collaboration and dialogue on a global scale to address the ethical, legal, and security challenges posed by autonomous weapons systems is paramount. By fostering transparency, accountability, and responsible innovation, the international community can work towards harnessing the benefits of autonomous weapons systems while mitigating the associated risks to humanity.

International Agreements on AI Weaponization

The international community is facing a pressing need to address the issue of AI weaponization through global agreements. As artificial intelligence continues to advance, the potential for autonomous weapons to be developed and deployed poses significant ethical and security concerns. International agreements are crucial to establish clear guidelines and regulations on the use of AI in weaponry to prevent unintended consequences and ensure accountability for any misuse. By fostering cooperation and establishing common standards, these agreements can help mitigate the risks associated with AI weaponization and promote responsible development and deployment of such technologies on a global scale. One key aspect that international agreements on AI weaponization should address is the need for transparency and accountability in the development and deployment of autonomous weapons. As AI systems become more sophisticated and autonomous, it becomes increasingly challenging to predict their behavior in complex and dynamic environments. This lack of predictability raises concerns about the potential for AI-powered weapons to act independently and make decisions that may not align with human intentions. International agreements can provide a framework for ensuring that developers and users of AI weapons are held accountable for the actions of these systems, thereby promoting greater transparency and oversight in the use of such technologies. International agreements on AI weaponization should also focus on the establishment of clear protocols for the ethical use of autonomous weapons. The deployment of AI-powered weapons raises profound ethical questions regarding the inherent value of human life, the principles of proportionality and distinction in warfare, and the responsibility of individuals and states for the consequences of their actions. By setting ethical guidelines and standards, international agreements can help shape the development and deployment of autonomous weapons in a way that upholds human rights and promotes compliance with international humanitarian law. In doing so, these agreements can contribute to fostering a more ethical and principled approach to the use of AI in weaponry, while also helping to build trust and confidence among nations in the global community.

XX. QUANTUM COMPUTING IN AI

Many experts believe that quantum computing holds the key to significantly advancing artificial intelligence capabilities. Quantum computing utilizes quantum-mechanical phenomena, such as superposition and entanglement, to perform calculations at a much faster rate than classical computers. This increased computational speed could revolutionize the field of AI by enabling more complex and powerful algorithms to be developed. With quantum computing, AI systems could process massive amounts of data in real-time, leading to more accurate predictions and smarter decision-making processes. Quantum computing has the potential to enhance machine learning algorithms that underpin many AI applications. By leveraging quantum algorithms, AI models could be trained more efficiently and effectively, allowing for quicker optimization and deployment of AI solutions. This synergy between quantum computing and AI could unlock new possibilities across various industries, including healthcare, finance, and cybersecurity. As quantum computing continues to mature and become more accessible, the integration of quantum technologies with AI is expected to drive rapid innovation and transform the capabilities of intelligent systems. The combination of quantum computing and AI could address some of the current limitations and challenges faced by traditional AI approaches. Quantum AI could enable researchers to tackle problems that are currently intractable with classical computers, such as optimizing complex systems and performing high-dimensional data analysis. This convergence of quantum computing and AI represents a paradigm shift in the field of artificial intelligence, promising to unlock new levels of intelligence and problem-solving abilities that were previously thought impossible. As researchers continue to explore the potential of quantum computing in AI, exciting breakthroughs and innovations are expected to emerge, shaping the future of intelligent systems.

Quantum Computing Fundamentals

Advances in technology have paved the way for quantum computing to emerge as a revolutionary field in the realm of artificial intelligence. Unlike classical computers which operate on bits represented as either a 0 or 1, quantum computers use qubits that can exist in a superposition of states, allowing for parallel processing and exponential speedups in complex calculations. This fundamental difference in processing power opens up new possibilities for solving intricate problems that are out of reach for conventional computers. At the core of quantum computing is the principle of superposition, where qubits can exist in multiple states simultaneously until measured. This property allows quantum computers to perform calculations on a massive scale with unprecedented efficiency, making them well-suited for tackling challenges in cryptography, optimization, and drug discovery. Entanglement, another key feature of quantum mechanics, enables qubits to be correlated in such a way that the state of one qubit can instantly affect the state of another, providing a powerful tool for information processing and communication. The development of quantum algorithms, such as Shor's algorithm for factorization and Grover's algorithm for searching, has demonstrated the potential of quantum computing to revolutionize various industries and fields. As research and advancements in quantum technology continue to progress, the future holds promise for even greater breakthroughs in AI. From enhancing machine learning algorithms to accelerating complex simulations, the integration of quantum computing into AI systems is poised to usher in a new era of innovation and discovery, shaping the landscape of artificial intelligence for years to come.

Quantum Machine Learning Algorithms

Quantum machine learning algorithms represent an exciting frontier in the field of artificial intelligence, offering the potential for exponential advancements in computing power and data processing capabilities. By harnessing the principles of quantum mechanics, these algorithms can tackle complex problems that are beyond the capabilities of classical machine learning models. Quantum machine learning leverages the unique properties of quantum systems such as superposition and entanglement to perform computations at a speed that surpasses classical computers, paving the way for unprecedented breakthroughs in AI research and application. One of the key advantages of quantum machine learning algorithms is their ability to handle vast amounts of data and perform parallel computations efficiently. Traditional machine learning algorithms face limitations in processing large datasets due to the sequential nature of classical computing, but quantum algorithms can exploit quantum parallelism to simultaneously analyze multiple data points. This parallel processing capability enables quantum machine learning models to uncover intricate patterns and correlations in data that may otherwise remain hidden, leading to more accurate predictions and insights across various industries. Quantum machine learning holds the promise of enhancing the performance of existing AI technologies, such as deep learning and reinforcement learning. By integrating quantum computing principles into these frameworks, researchers can potentially accelerate training processes, optimize model efficiency, and expand the scope of AI applications. As quantum computing continues to advance, the synergy between quantum mechanics and machine learning is poised to revolutionize the landscape of AI, driving innovation in areas ranging from autonomous systems to advanced data analytics. The development of quantum machine learning algorithms represents a paradigm shift that could shape the future of artificial intelligence and propel us towards the next era of technological evolution.

Quantum Neural Networks

The advancement of artificial intelligence has taken a significant leap with the emergence of quantum neural networks. These networks operate by harnessing the principles of quantum mechanics to process information at a level far beyond traditional computing methods. Unlike classical neural networks, which rely on binary data processing, quantum neural networks leverage the complex phenomena of superposition and entanglement to perform calculations exponentially faster. This revolutionary approach has the potential to revolutionize the field of artificial intelligence by enabling the development of more powerful and efficient learning models. One of the key advantages of quantum neural networks is their ability to handle vast amounts of data and complex computations in a fraction of the time required by classical computers. This opens up new possibilities for solving

114

highly intricate problems in fields such as cryptography, optimization, and pattern recognition. By exploiting the inherent properties of quantum physics, these networks can achieve unprecedented levels of accuracy and precision in data analysis and decision-making processes. As a result, quantum neural networks have the potential to drive major breakthroughs in AI research and applications, paving the way for more sophisticated intelligent systems. Despite the immense potential of quantum neural networks, their development is still in its early stages, facing challenges such as scalability, error rates, and the need for specialized hardware. Ongoing research and advancements in quantum computing technology are steadily overcoming these obstacles, bringing us closer to realizing the full potential of this cutting-edge approach to artificial intelligence. As quantum neural networks continue to evolve and mature, they have the potential to redefine the capabilities of AI systems, opening up new frontiers in innovation and revolutionizing the way we interact with technology in the future.

XXI. EXPLAINABLE AI IN AUTONOMOUS SYSTEMS

As autonomous systems become increasingly prevalent in various industries, the concept of explainable AI has gained significant importance. Explainable AI focuses on making the decision-making process of autonomous systems transparent and understandable to humans. Through transparent algorithms and clear logic, autonomous robots can provide explanations for their actions, enhancing trust and acceptance among users. This is particularly crucial in applications where human lives are at stake, such as autonomous vehicles or medical robots. One of the main challenges in achieving explainable AI in autonomous systems lies in balancing the need for transparency with the complexity of the algorithms involved. As autonomous robots rely on intricate machine learning models and neural networks to make decisions, translating these processes into understandable explanations can be a daunting task. Researchers are exploring innovative techniques, such as interpretable machine learning models and interactive visualization tools, to enhance the interpretability of autonomous systems without compromising their performance. Explainable AI in autonomous systems is not only vital for building trust with users but also for ensuring accountability and ethics in decision-making processes. By providing clear explanations for their actions, autonomous robots can enable humans to intervene when needed and prevent potentially harmful outcomes. Implementing explainable AI in autonomous systems is a significant step towards creating responsible and reliable technologies that can coexist harmoniously with humans in various domains.

Importance of Explainable AI in Autonomous Robots

The importance of explainable AI in autonomous robots cannot be overstated in the current technological landscape. As these robots are designed to operate independently in various environments, it is crucial for users to understand how they make decisions and the reasoning behind their actions. Having transparency in the decision-making process of autonomous robots is essential for building trust with users and ensuring that their actions align with ethical and moral standards. Explainable AI enables humans to comprehend the rationale behind the choices made by these robots, leading to better acceptance and cooperation between man and machine. Explainable AI plays a critical role in error detection and troubleshooting in autonomous robots. By providing insights into the decision-making process, engineers and developers can identify potential issues or biases that may arise in the robot's behavior. This transparency allows for timely intervention and improvement of the system, leading to more reliable and efficient performance. Explainable AI helps in enhancing the overall safety and security of autonomous robots by enabling operators to understand the root cause of any malfunctions or errors that occur during operation. Explainable AI aids in regulatory compliance and accountability in the deployment of autonomous robots. With clear explanations of how decisions are made, it becomes easier to ensure that these robots adhere to legal and ethical standards in different industries. By promoting transparency and accountability, explainable AI can mitigate potential risks and negative impacts that autonomous robots may pose to society. Integrating explainable AI into autonomous robots is crucial for fostering trust, improving performance, ensuring safety, and upholding ethical standards in the development and deployment of these advanced technologies.

Techniques for Interpretable AI Models in Robotics

As the field of robotics continues to advance, the demand for interpretable AI models becomes increasingly crucial. Techniques such as explainable artificial intelligence (XAI) provide insights into how AI models make decisions, enhancing transparency and trust in autonomous systems. By using visualization tools, researchers can interpret complex deep learning models, making it easier to understand the reasoning behind machine decisions. This interpretability not only improves the reliability of robots but also ensures their safe integration into various domains, such as healthcare, manufacturing, and transportation. Another effective technique for creating interpretable AI models in robotics is the use of rule-based systems. These systems rely on predefined rules that govern the behavior of robots, allowing designers to encode human knowledge and expertise into the decision-making process. By combining symbolic reasoning with machine learning, rule-based approaches enable robots to explain their actions in a clear and easily understandable manner. This transparency is crucial in scenarios where human operators need to understand and trust the decisions made by autonomous systems, ensuring effective collaboration between humans and robots in complex environments. Incorporating human feedback into the training process of AI models can significantly enhance interpretability in robotics. By involving humans in the loop, researchers can gather insights into the decision-making process of robots and refine their algorithms based on real-world feedback. This iterative approach not only improves the performance of AI models but also increases their interpretability by incorporating human intuition and context. As robots become more integrated into daily life, the ability to understand and interpret machine behavior becomes essential for ensuring a seamless interaction between humans and autonomous systems.

Applications in Critical Decision-Making Systems for Autonomous Robots

As autonomous robots continue to push the boundaries of artificial intelligence, they offer a multitude of applications in critical decision-making systems. These sophisticated machines are designed to operate independently, using complex algorithms and sensory inputs to navigate and interact with their environment. In sectors like healthcare, autonomous robots can assist in surgeries, drug delivery, and patient care, making real-time decisions based on data analysis and machine learning. In logistics, these robots can optimize warehouse operations, inventory management, and delivery services, enhancing efficiency and reducing human error. The ability of autonomous robots to make critical decisions in dynamic and unpredictable environments is reshaping industries such as agriculture, security, and manufacturing. In agriculture, they can autonomously plant and harvest crops, monitor soil conditions, and detect pests, revolutionizing farming practices and maximizing crop yields. In the security sector, autonomous robots can patrol high-risk areas, detect intruders, and respond to emergencies in a timely manner, enhancing public safety and reducing human risk. In manufacturing, these robots can streamline production processes, perform repetitive tasks with precision, and adapt to changes in production demands, increasing productivity and reducing operational costs. As autonomous robots become more integrated into our daily lives and industries, ethical considerations and technological challenges arise. Issues such as data privacy, security vulnerabilities, and job displacement need to be carefully addressed to ensure the responsible development and deployment of autonomous robots. It is crucial for policymakers, technologists, and ethicists to collaborate in establishing regulations and ethical frameworks that promote the safe and ethical use of autonomous robots while maximizing their potential to improve decision-making processes in various sectors.

XXII. AI IN AGRICULTURE

Advancements in artificial intelligence have paved the way for transformative applications in various industries, including agriculture. The integration of AI technologies in agriculture has revolutionized traditional farming practices, enhancing efficiency, productivity, and sustainability. One of the key areas where AI has had a significant impact is in crop monitoring and management. Through the use of sensors, drones, and satellite imagery, AI algorithms can analyze data on crop health, soil conditions, and weather patterns to provide real-time insights to farmers. This enables farmers to make data-driven decisions to optimize crop yields and minimize resource wastage. In addition to crop monitoring, AI-powered solutions are also being used for precision agriculture, enabling farmers to deploy targeted treatments based on precise data analysis. Machine learning algorithms can analyze vast amounts of data to predict crop yields, detect diseases early, and optimize irrigation schedules. This not only improves the overall efficiency of farming operations but also reduces the environmental impact by minimizing the use of water, pesticides, and fertilizers. AI-driven robotic systems are being developed to automate tasks such as planting, weeding, and harvesting, reducing the need for manual labor and increasing the scalability of agricultural operations. The integration of AI in agriculture holds great potential to address the challenges of food security, climate change, and resource scarcity. By leveraging the power of AI technologies, farmers can optimize their operations, increase sustainability, and ensure a more reliable food supply for the growing global population. The widespread adoption of AI in agriculture also raises questions about data privacy, ethics, and the potential displacement of labor. It is crucial for policymakers, researchers, and industry stakeholders to collaborate and address these challenges to maximize the benefits of AI while mitigating potential risks.

Precision Agriculture with AI

As artificial intelligence continues to advance, precision agriculture stands to benefit significantly from the integration of AI technologies. AI can revolutionize the way farmers analyze data, optimize resources, and make informed decisions to enhance agricultural productivity. By utilizing AI algorithms and machine learning models, farmers can collect and process vast amounts of data from sensors, drones, and satellites to monitor crop health, predict yields, and apply fertilizers and pesticides more efficiently. This level of precision allows farmers to minimize waste, reduce environmental impact, and increase overall crop yields, ultimately leading to a more sustainable and productive agricultural sector. AI-powered precision agriculture can enable farmers to implement real-time monitoring and automated decision-making processes, reducing the need for manual labor and increasing operational efficiency. Autonomous farming vehicles equipped with AI technology can autonomously navigate fields, perform tasks such as planting, harvesting, and spraying, and even detect and address crop diseases or pests without human intervention. This level of autonomy not only saves time and labor costs for farmers but also ensures timely and precise actions to maximize crop yield and quality. AI can help farmers in predicting weather patterns, market trends, and disease outbreaks, allowing them to plan and mitigate risks effectively. The integration of AI in precision agriculture represents a significant step towards sustainable and efficient farming practices. By leveraging AI technologies, farmers can optimize resource allocation, reduce waste, and improve crop yields while also minimizing environmental impact. As AI continues to evolve, the possibilities for enhancing precision agriculture are endless, offering the potential for increased food security, economic growth, and environmental sustainability in the agricultural sector. Through continued research, innovation, and adoption of AI solutions, the future of precision agriculture with AI looks promising, paving the way for a more efficient and sustainable approach to farming in the years to come.

AI-driven Crop Monitoring and Management

As technology continues to advance, artificial intelligence has made significant strides in various fields, including agriculture. AI-driven crop monitoring and management is a revolutionary approach that leverages cutting-edge technologies to optimize farming practices and increase crop yield. By utilizing sophisticated algorithms and machine learning models, farmers can now monitor crop health, detect diseases, predict yields, and make data-driven decisions in real-time. This level of precision and efficiency has the potential to revolutionize traditional farming methods and improve overall agricultural productivity. One key aspect of AI-driven crop monitoring and management is the ability to collect and analyze vast amounts of data from multiple sources, such as satellite imagery, drones, weather stations, and sensors. This data is then processed using advanced analytical tools to provide valuable insights into crop conditions, soil health, water usage, and pest infestations. By harnessing the power of AI, farmers can access actionable information that

123

enables them to optimize resource allocation, reduce waste, and mitigate risks effectively. This proactive approach not only enhances crop quality and yield but also contributes to sustainable farming practices by minimizing environmental impact. AI-driven crop monitoring and management systems can help farmers make informed decisions that align with market demands and consumer preferences. By analyzing market trends, consumer behavior, and supply chain dynamics, AI technologies can provide valuable insights into pricing strategies, distribution channels, and product positioning. This strategic advantage enables farmers to adapt quickly to changing market conditions, capitalize on emerging opportunities, and optimize revenue generation. AI-driven crop monitoring and management represents a paradigm shift in agriculture, empowering farmers with the tools and knowledge needed to navigate an increasingly complex and competitive industry landscape.

Ethical Considerations in AI Applications for Sustainable Agriculture

In the realm of sustainable agriculture, the integration of artificial intelligence applications represents a promising avenue for addressing key challenges such as resource management, crop optimization, and yield prediction. The adoption of AI in agriculture raises important ethical considerations that must be carefully addressed to ensure the technology's responsible and sustainable use. One ethical concern revolves around data privacy and ownership, as AI systems rely heavily on vast amounts of data collected from farmers and agricultural operations. Safeguarding this data from unauthorized access and misuse is paramount to maintaining the trust and confidence of stake-holders within the agricultural ecosystem. Another ethical consideration in the deployment of AI in sustainable agriculture is the potential for exacerbating social inequalities. As AI technologies increasingly automate various tasks and decision-making processes on farms, there is a risk of widening the digital divide between large-scale commercial farms with access to advanced AI tools and smaller, resource-constrained farms that may be left behind. Ensuring equitable access to AI technologies and supporting capacity-building initiatives for farmers of all scales can help mitigate these disparities and promote inclusive and sustainable agricultural practices. Transparency and accountability in the development and implementation of AI systems in agriculture are essential to ensure that decisions made by these technologies align with ethical standards and societal values. The responsible use of AI in sustainable agriculture necessitates ongoing dialogue and collaboration between developers, policymakers, farmers, and other stakeholders to establish clear guidelines and frameworks for ethical AI deployment. This collaborative approach can help address emerging ethical challenges, such as algorithmic bias, interpretability of AI-driven recommendations, and the overall impact of AI on rural communities and ecosystems. By fostering a culture of ethical awareness and responsibility in AI applications for sustainable agriculture, we can harness the transformative potential of technology while safeguarding the well-being of farmers, consumers, and the environment.

XXIII. AI IN TRANSPORTATION

One significant domain where artificial intelligence has made a profound impact is in transportation. The integration of AI technologies in transportation systems has revolutionized the way we commute and transport goods. Autonomous vehicles, guided by AI algorithms, have the potential to enhance road safety, reduce traffic congestion, and optimize fuel efficiency. Through real-time data analysis and machine learning, AI enables vehicles to make split-second decisions, leading to more efficient and reliable transportation networks. The use of AI in transportation is not limited to autonomous vehicles but also extends to traffic management systems, logistics planning, and predictive maintenance. AI-powered solutions can analyze vast amounts of data to predict traffic patterns, optimize routes, and schedule maintenance activities proactively. By harnessing the power of AI, transportation companies can streamline operations, reduce costs, and improve customer satisfaction. AI technologies pave the way for the development of smart cities, where interconnected transportation systems facilitate seamless and efficient urban mobility. Despite the tremendous benefits that AI brings to the transportation sector, challenges such as data privacy, cybersecurity, and regulatory frameworks must be addressed to ensure the safe and ethical use of AI technologies. As autonomous vehicles become more prevalent on our roads, policymakers, businesses, and the public must collaborate to establish guidelines and standards that prioritize safety, security, and accountability. By navigating these challenges thoughtfully, the integration of AI in transportation has the potential to revolutionize the way we move people and goods, leading to a more sustainable, efficient, and interconnected transportation ecosystem.

Autonomous Vehicles and AI

Advances in artificial intelligence have paved the way for the development of autonomous vehicles, a revolutionary technology that promises to transform the transportation industry. Autonomous vehicles, also known as self-driving cars, leverage AI algorithms to perceive their surroundings, navigate roads, and make real-time decisions without human intervention. These vehicles rely on a combination of sensors, cameras, radar, and Lidar technology to detect obstacles, interpret traffic signs, and predict the movements of other vehicles on the road. By integrating AI into autonomous vehicles, manufacturers aim to improve road safety, reduce traffic congestion, and enhance the overall mobility experience for passengers. One of the key challenges in the development of autonomous vehicles is ensuring the robustness and reliability of AI algorithms in complex and dynamic driving environments. Autonomous vehicles must be equipped with AI systems that can process vast amounts of data in real-time, make split-second decisions, and adapt to unforeseen situations on the road. This requires continuous testing, validation, and refinement of AI models to ensure that autonomous vehicles can operate safely and efficiently in diverse driving conditions. Ethical considerations surrounding autonomous vehicles, such as liability in the event of accidents and the impact on employment in the transportation sector, must be carefully addressed to gain public acceptance and trust. Despite these challenges, the potential benefits of autonomous vehicles powered by AI are immense. From reducing traffic accidents and fatalities to improving traffic flow and reducing emissions, autonomous vehicles have the potential to revolutionize the way we travel. As AI technologies continue to advance and regulatory frameworks evolve to address safety and ethical concerns, autonomous vehicles are poised to become an integral part of the future transportation ecosystem. By harnessing the power of AI, autonomous vehicles have the potential to make urban mobility more sustainable, efficient, and accessible for people around the world.

Traffic Management Systems with AI

Traffic management systems incorporating artificial intelligence have emerged as a promising solution to improve road safety, reduce congestion, and enhance overall transportation efficiency. By utilizing AI algorithms and machine learning, these systems can analyze real-time traffic data, predict traffic patterns, and optimize traffic flow in a dynamic and responsive manner. Through the integration of sensors, cameras, and connected devices, AI-powered traffic management systems can adapt to changing traffic conditions, prioritize emergency vehicles, and facilitate smoother coordination between different modes of transportation. One key advantage of traffic management systems with AI is their ability to not only react to traffic incidents but also proactively predict and prevent potential bottlenecks or accidents. By leveraging data analytics and predictive modeling, these systems can identify patterns and trends in traffic behavior, allowing for early intervention strategies to be implemented. AI algorithms can learn from historical data and continuously improve their decision-making processes, leading to more accurate and efficient traffic management strategies over time. This adaptive and self-learning capability sets AI-powered traffic management systems apart from traditional rule-based systems, making them well-suited for the complexities of modern urban environments. In addition to optimizing traffic flow and improving road safety, AI-powered traffic management systems can also contribute to reducing emissions and enhancing environmental sustainability. By optimizing traffic signals, route planning, and vehicle speeds, these systems can help minimize fuel consumption, lower carbon emissions, and promote eco-friendly transportation practices. By facilitating the integration of electric vehicles and other alternative transportation modes, AI-powered traffic management systems can support the transition towards a more sustainable and greener urban transportation ecosystem. The combination of AI technologies with traffic management systems holds great promise in transforming the way we plan, manage, and experience traffic in our cities.

Safety and Ethical Concerns in AI-driven Transportation

One of the primary concerns surrounding the integration of AI in transportation is the issue of safety. As autonomous vehicles and other AI-driven transportation systems become more prevalent, ensuring the safety of passengers, pedestrians, and other road users is of paramount importance. The complex decision-making processes required for safe navigation in unpredictable environments pose a significant challenge for AI systems. The potential for system malfunctions, hacking, or external interference raises concerns about the reliability of AI-driven transportation modes. These safety concerns must be addressed through rigorous testing, regulation, and continuous monitoring of AI systems to minimize the risks associated with their deployment on a large scale. In addition to safety considerations, ethical concerns play a crucial role in shaping the future of AI-driven transportation. The ethical implications of AI decision-making, particularly in life-or-death situations, raise questions about accountability, transparency, and the moral values embedded in AI algorithms. Issues such as the prioritization of passenger safety over pedestrians, the impact of AI biases on decision-making processes, and the potential for human-AI conflicts in critical situations highlight the need for robust ethical frameworks to guide the development and deployment of AI in transportation. Striking a balance between technological advancement and ethical considerations is essential to ensure that AI-driven transportation systems align with societal values and norms. Addressing safety and ethical concerns in AI-driven transportation requires a multidisciplinary approach that encompasses technical expertise, ethical reasoning, and regulatory frameworks. Collaboration between engineers, ethicists, policymakers, and other stakeholders is essential to develop comprehensive guidelines for the design, operation, and regulation of AI systems in transportation. Transparency in AI decision-making processes, accountability mechanisms for system failures, and proactive measures to minimize risks are crucial steps towards building trust in AI-driven transportation systems.

XXIV. AI IN RETAIL

The integration of artificial intelligence in the retail sector has revolutionized the way businesses engage with customers and optimize their operations. One key application of AI in retail is personalized shopping experiences. By leveraging algorithms and machine learning, retailers can analyze vast amounts of customer data to tailor product recommendations and promotions to individual preferences. This not only enhances customer satisfaction but also boosts sales and customer loyalty. As AI continues to evolve, retailers are also exploring the use of chatbots and virtual assistants to provide real-time customer support, helping to streamline the shopping experience and improve overall customer satisfaction. AI-powered analytics tools are enabling retailers to make data-driven decisions and optimize various aspects of their business, from inventory management to supply chain operations. By analyzing trends and patterns in sales data, retailers can forecast demand more accurately, reduce stockouts, and improve inventory turnover. AI can help retailers identify fraud and prevent potential losses through advanced fraud detection algorithms. By automating routine tasks and processes, AI not only frees up employees to focus on higher-value activities but also improves operational efficiency and reduces costs for retailers. Looking ahead, the adoption of AI in retail is expected to continue at a rapid pace, with advancements in technologies like computer vision and natural language processing further enhancing the capabilities of AI-powered solutions. Retailers are also exploring the potential of autonomous robots for tasks such as inventory management, restocking shelves, and even cashier-less checkout processes.

While the integration of AI in retail presents numerous opportunities for innovation and growth, it also raises ethical considerations around data privacy, security, and job displacement. As the retail industry embraces AI, it will be essential for businesses to navigate these challenges thoughtfully and responsibly to ensure a successful and sustainable future for AI in retail.

Personalized Shopping Experiences with AI

The exponential growth of artificial intelligence has revolutionized the way businesses deliver personalized shopping experiences to consumers. Through the utilization of AI-powered technologies, companies can now analyze vast amounts of data to tailor products and services to individual preferences and behavior patterns. By harnessing the power of machine learning and predictive analytics, businesses can offer recommendations, discounts, and promotions tailored specifically to each customer. This level of personalization not only enhances the overall shopping experience but also fosters stronger customer loyalty and increases revenue potential. AI has enabled the development of virtual shopping assistants that can provide real-time assistance to customers, guiding them through their shopping journey and answering any questions they may have. These virtual assistants utilize natural language processing to understand and respond to customer queries effectively. By bridging the gap between online and offline shopping experiences, virtual assistants enhance customer satisfaction and drive engagement. The integration of chatbots powered by AI technology allows for efficient and personalized communication with customers, further enhancing the shopping experience. The evolution of AI in the retail sector has paved the way for the implementation of autonomous robots in physical stores and warehouses. These robots can assist in inventory management, restocking shelves, and even providing customer service. By automating routine tasks, businesses can streamline operations, reduce costs, and improve overall efficiency. The integration of autonomous robots also raises ethical considerations regarding potential job displacement and the need for regulations to ensure the safe and ethical use of AI technology in retail settings. AI has transformed personalized shopping experiences by offering tailored product recommendations, virtual shopping assistants, and autonomous robots, revolutionizing the way businesses interact with customers and enhancing overall efficiency in the retail sector.

Inventory Management and Supply Chain Optimization

As businesses strive to optimize their supply chain efficiency, inventory management plays a crucial role in achieving this objective. By carefully monitoring and controlling the flow of goods, companies can minimize excess inventory costs, reduce stockouts, and improve overall customer satisfaction. Utilizing advanced technologies such as artificial intelligence can significantly enhance inventory management processes by providing real-time insights into demand forecasting, inventory levels, and replenishment strategies. AI-powered systems can analyze vast amounts of data quickly and accurately, enabling companies to make informed decisions that drive operational excellence and profitability. Integrating AI into inventory management allows for predictive analytics that can identify potential bottlenecks or disruptions in the supply chain before they occur. By leveraging machine learning algorithms, organizations can anticipate demand fluctuations, optimize reorder points, and streamline inventory turnover rates. This proactive approach not only enhances operational efficiency but also minimizes the risk of excess stock accumulation or shortages, ultimately leading to cost savings and improved customer service levels. AI can automate routine inventory tasks such as order processing, inventory tracking, and demand forecasting, freeing up human resources to focus on more strategic initiatives that drive business growth and innovation. The strategic implementation of AI in inventory management and supply chain optimization holds immense potential for businesses looking to gain a competitive edge in today's fast-paced global marketplace. By harnessing the power of artificial intelligence to forecast demand, optimize inventory levels, and enhance supply chain visibility, companies can improve resource allocation, reduce operational costs, and deliver superior customer experiences. As technology continues to evolve, organizations that embrace AI-driven solutions in their inventory management processes will be better positioned to adapt to changing market dynamics and achieve sustainable growth in the digital age.

Privacy and Data Security in AI-powered Retail Solutions

A major concern surrounding AI-powered retail solutions is the issue of privacy and data security. As these systems become more integrated into our daily lives, the vast amount of personal data they collect raises questions about how this information is being used and protected. Retailers rely on AI algorithms to analyze consumer behavior, preferences, and purchasing patterns to deliver personalized recommendations and targeted advertisements. This level of data collection can be invasive and raises concerns about user privacy. The risk of data breaches and unauthorized access to sensitive information poses a significant threat to consumers and the reputation of the companies utilizing these technologies. In the context of AI-powered retail solutions, ensuring robust data security measures is crucial to maintaining customer trust and complying with legal requirements. Retailers must implement encryption techniques, secure data storage protocols, and stringent access controls to protect consumer data from cyber threats. Transparency in data collection practices and providing clear opt-out options can help mitigate privacy concerns and empower consumers to make informed decisions about sharing their information. By prioritizing data security and privacy in the design and implementation of AI systems, retailers can build a foundation of trust with their customers and foster long-term relationships based on mutual respect and transparency. Despite the challenges and risks associated with privacy and data security in AI-powered retail solutions, there are opportunities for innovation and advancement in the field. By investing in research and development of privacy-enhancing technologies such as differential privacy and federated learning, companies can strike a balance between leveraging consumer data for business insights and protecting individual privacy rights. Collaborating with regulators, privacy advocates, and cybersecurity experts can also help retailers navigate the complex landscape of data protection laws and industry best practices.

XXV. AI IN SOCIAL MEDIA

Advances in artificial intelligence have revolutionized the way we interact with social media platforms. AI-powered algorithms are utilized to analyze user data and behavior, leading to personalized content recommendations and targeted advertising. These algorithms can sift through massive amounts of data to extract valuable insights, enabling businesses to optimize their marketing strategies and enhance user engagement. AI has facilitated the development of chatbots that provide instant customer support on social media platforms, improving user experience and efficiency. The integration of AI in social media has not only transformed how businesses connect with their target audience but has also enhanced the overall user experience by tailoring content to individual preferences. AI has played a crucial role in combating fake news and misinformation on social media platforms. By leveraging machine learning algorithms, AI can detect and flag suspicious content, helping to maintain the credibility and integrity of online information sources. AI-powered content moderation tools can automatically filter out inappropriate or offensive material, safeguarding users from harmful content. This proactive approach to content moderation enhances user safety and fosters a more positive online environment. As the volume of digital content continues to grow exponentially, the ability of AI to sift through and curate information effectively becomes increasingly essential in maintaining the quality and reliability of social media platforms. Looking forward, the evolution of AI in social media is expected to further impact the way businesses and individuals interact online. As AI continues to advance, there is the potential for even more sophisticated algorithms to personalize content and tailor user experiences. The integration of AI with augmented reality and virtual reality technologies opens up new possibilities for immersive and interactive social media experiences. As AI becomes more prevalent in the realm of social media, ethical considerations regarding data privacy, algorithmic bias, and content censorship must be carefully navigated to ensure a fair and transparent online environment for all users. The fusion of AI with social media is reshaping the digital landscape, offering both opportunities for innovation and challenges for ethical governance.

AI for Content Recommendation

The advancement of artificial intelligence has brought about a significant shift in the way content is recommended to users. AI-driven content recommendation systems have revolutionized the digital landscape by providing personalized suggestions based on users' preferences, behavior, and interactions. These systems leverage sophisticated algorithms, machine learning models, and natural language processing techniques to analyze vast amounts of data and accurately predict the content that will best resonate with individual users. As a result, users are presented with content that is tailored to their interests, leading to enhanced user experience and increased engagement. One of the key benefits of AI-powered content recommendation systems is their ability to continuously learn and adapt to user preferences in real-time. By analyzing user interactions, feedback, and behavior patterns, these systems can dynamically adjust their recommendations to ensure that users are presented with content that is most relevant and engaging to them. This personalized approach enhances user satisfaction, increases user retention, and ultimately drives revenue by promoting longer user sessions and higher conversion rates. AI-based content recommendation systems can help content providers and marketers better understand their audience, allowing them to fine-tune their content strategies and optimize their offerings for maximum impact. Moving forward, the future of AI for content recommendation holds promising opportunities for further innovation and refinement. As technology continues to evolve, we can expect to see even more sophisticated AI models, enhanced data analysis capabilities, and seamless integration across different platforms and devices. With the continued advancement of AI, content recommendation systems will become even more personalized, intuitive, and user-centric, driving deeper engagement and delivering more value to both users and content providers. It is crucial to address ethical considerations, such as data privacy and transparency, to ensure that AI-powered content recommendation systems operate ethically and responsibly in a rapidly evolving digital landscape.

Sentiment Analysis and Social Listening

As artificial intelligence continues to advance, sentiment analysis and social listening have emerged as crucial components of understanding human behavior and interactions in the digital realm. Sentiment analysis refers to the process of computationally identifying and categorizing opinions, emotions, and attitudes expressed in text data, while social listening involves monitoring online conversations to gather insights and trends. By analyzing social media posts, customer reviews, and other textual data, businesses can gauge public sentiment, identify customer preferences, and tailor their marketing strategies accordingly. This not only helps companies improve their products and services but also enhances customer satisfaction and loyalty. The integration of sentiment analysis and social listening into AI systems has revolutionized the way businesses engage with their audience and make data-driven decisions. Through sentiment analysis, organizations can quickly assess the overall sentiment surrounding their brand, products, or services, allowing them to address any negative feedback or concerns promptly. Social listening provides valuable insights into consumer behavior, market trends, and competitor strategies, enabling companies to stay ahead of the curve and adapt to changing market dynamics. This proactive approach to monitoring and analyzing online conversations gives businesses a competitive edge in understanding their target audience and optimizing their marketing efforts effectively. The combination of sentiment analysis and social listening has broader implications beyond business applications. Researchers and policymakers can use these tools to analyze public opinion on social and political issues, track trends in public sentiment, and even predict potential societal shifts. By harnessing the power of AI to analyze vast amounts of textual data, valuable insights can be gained into the collective mindset of society, helping to inform decision-making processes, shape policies, and improve overall social welfare. Sentiment analysis and social listening represent an invaluable resource in leveraging AI to gain deeper insights into human behavior and interactions in the age of digital communication.

Ethical Implications of AI in Social Media Platforms

As artificial intelligence continues to advance, its integration into social media platforms has raised significant ethical implications. One of the primary concerns is the potential manipulation of user data for targeted advertising or political purposes. AI algorithms can analyze vast amounts of personal information to create profiles that can be exploited for profit or influence. This raises questions about the privacy and autonomy of individuals online, as their behaviors and preferences may be used without their full understanding or consent. The opaque nature of AI decision-making processes can lead to biases and discrimination, further exacerbating existing social inequalities. Another ethical dilemma presented by AI in social media platforms is the spread of misinformation and fake news. AI-powered algorithms can amplify the dissemination of false information, leading to widespread confusion and discord within online communities. The challenge lies in balancing the need for freedom of expression with the responsibility to uphold accuracy and integrity in information sharing. As social media becomes a primary source of news and information for many individuals, the ethical implications of AI in perpetuating falsehoods and promoting divisiveness cannot be underestimated. The influence of AI in shaping online user experiences raises concerns about the erosion of human agency and autonomy. By personalizing content and recommendations based on predictive algorithms, social media platforms can create filter bubbles that reinforce existing beliefs and preferences, limiting exposure to diverse perspectives. This can result in echo chambers that hinder critical thinking and open discourse, ultimately undermining the democratic ideals of a free and informed society. In light of these ethical implications, it is imperative for regulators, policymakers, and technology companies to prioritize transparency, accountability, and user empowerment in the deployment of AI in social media platforms.

XXVI. AI IN LAW ENFORCEMENT

The integration of artificial intelligence in law enforcement has raised important ethical and practical considerations. The use of AI technologies, such as facial recognition software, predictive policing algorithms, and autonomous drones, has significantly impacted how law enforcement agencies operate. While these technologies offer the promise of increased efficiency and effectiveness in crime prevention and investigation, they also raise concerns about privacy, bias, and accountability. The deployment of AI in law enforcement requires careful regulation and oversight to ensure that these technologies are used ethically and responsibly. One of the key benefits of AI in law enforcement is its potential to enhance public safety by enabling proactive crime prevention strategies. Predictive policing algorithms, for example, can analyze vast amounts of data to identify patterns and trends that may indicate where crimes are likely to occur. This data-driven approach allows law enforcement agencies to allocate resources more effectively and respond to potential threats before they escalate. The use of AI in predictive policing has also been criticized for perpetuating existing biases in the criminal justice system and unfairly targeting marginalized communities. It is essential for policymakers and law enforcement agencies to address these concerns and implement safeguards to prevent the misuse of AI technologies in policing. Another important aspect of AI in law enforcement is the use of facial recognition technology for identifying and tracking individuals. While facial recognition can be a powerful tool for identifying suspects and enhancing security, it also raises serious privacy and civil liberties concerns. The indiscrimulate use of facial recognition software by law enforcement agencies poses a significant risk to individuals' privacy rights and may lead to wrongful arrests or discriminatory targeting. Regulations must be put in place to ensure that facial recognition technology is used lawfully and transparently, with appropriate safeguards to protect individuals' rights and prevent abuse. The integration of AI in law enforcement requires a delicate balance between leveraging the benefits of technology for public safety and upholding the principles of fairness, accountability, and respect for civil liberties.

Predictive Policing with AI

As artificial intelligence continues to advance, the integration of predictive policing with AI has shown great potential in enhancing law enforcement strategies. By utilizing vast amounts of data and sophisticated algorithms, predictive policing aims to forecast potential criminal activities and optimize resource allocation for crime prevention. This proactive approach allows law enforcement agencies to anticipate and prevent crimes before they occur, ultimately leading to a safer community. The predictive models can analyze historical crime data, weather patterns, social media activity, and other relevant factors to identify high-risk areas and individuals, enabling law enforcement to focus their efforts on crime hotspots and high-risk individuals. Predictive policing with AI has the potential to address biases present in traditional policing methods. By relying on data-driven insights and algorithms, rather than subjective human judgment, predictive policing can help reduce racial profiling and discrimination in law enforcement practices. The algorithms can identify patterns and trends objectively, leading to more equitable and effective decision-making processes. It is crucial to continuously monitor and evaluate these predictive models to ensure they are fair and unbiased. Transparency and accountability in the development and implementation of AI algorithms are essential to mitigate potential risks and uphold ethical standards in predictive policing practices. The integration of AI in predictive policing represents a significant advancement in law enforcement capabilities. By harnessing the power of data analytics and machine learning, law enforcement agencies can enhance their crime prevention strategies, allocate resources more efficiently, and reduce biases in policing practices. As the technology continues to evolve, it is essential for policymakers, law enforcement agencies, and AI developers to collaborate in establishing clear guidelines and regulations to ensure the responsible and ethical use of predictive policing with AI. By striking a balance between technological innovation and ethical considerations, society can harness the full potential of AI in creating safer and more resilient communities.

Facial Recognition Technology in Law Enforcement

The use of facial recognition technology in law enforcement has sparked significant debate regarding its ethical implications and potential for abuse. While proponents argue that such technology can enhance public safety by helping to identify and track suspects more efficiently, critics raise concerns about invasion of privacy, racial bias, and lack of transparency in its implementation. The widespread adoption of facial recognition technology by law enforcement agencies raises important questions about the balance between security and civil liberties, as well as the need for clear guidelines and oversight to prevent misuse. One of the key issues surrounding facial recognition technology in law enforcement is the accuracy and reliability of the algorithms used to identify individuals. Studies have shown that these algorithms can produce false matches, particularly for individuals of color, leading to wrongful arrests and a perpetuation of systemic biases. The lack of regulation and standardization in the development and deployment of facial recognition technology can exacerbate these disparities, highlighting the need for greater transparency and accountability in its use by law enforcement agencies. In light of these concerns, it is crucial for policymakers, technologists, and civil society to engage in a dialogue about the ethical and legal implications of facial recognition technology in law enforcement. This dialogue should include a consideration of the potential impact on marginalized communities, the need for robust privacy protections, and the development of clear guidelines for its use. By addressing these issues proactively, society can harness the benefits of facial recognition technology while mitigating its risks and ensuring that it is deployed ethically and responsibly.

Legal and Ethical Challenges in AI-assisted Crime Prevention

One of the main challenges in the development and implementation of AI-assisted crime prevention systems is the legal and ethical considerations that arise. As AI technologies become more sophisticated and autonomous, questions surrounding accountability and transparency become increasingly pertinent. Who will be held responsible if an AI system makes a mistake or misidentifies a suspect? How do we ensure that these systems are not biased or discriminatory in their decision-making processes? These legal and ethical dilemmas must be carefully addressed to prevent potential harm or injustice resulting from the use of AI in crime prevention. Issues related to privacy and data protection also come to the forefront when discussing AI-assisted crime prevention. The vast amount of data needed to train AI algorithms for crime detection and prediction raises concerns about surveillance and potential breaches of privacy. As these systems collect and analyze personal information, there is a risk of infringing on individuals' rights to privacy and autonomy. Striking a balance between using AI to enhance crime prevention efforts and safeguarding individuals' privacy rights presents a complex ethical challenge that must be navigated in a careful and deliberate manner. In addition to legal and ethical considerations, there is also a need to address the potential for misuse or abuse of AI-assisted crime prevention tools. As these technologies become more mainstream, there is a risk that they could be weaponized for malicious purposes, such as surveillance of political dissidents or the manipulation of evidence in criminal investigations. It is essential to establish robust regulations and safeguards to prevent the misuse of AI in the context of crime prevention and ensure that these technologies are deployed in a responsible and ethical manner that upholds the principles of justice and fairness.

XXVII. AI IN SPACE EXPLORATION

Advances in artificial intelligence have revolutionized numerous industries, including space exploration. AI technology has enabled researchers and scientists to analyze massive amounts of data collected from space missions more efficiently and effectively than ever before. Machine learning algorithms can process and interpret complex data sets, leading to new insights and discoveries that would have been nearly impossible to uncover through traditional methods alone. This has resulted in a significant acceleration of research and experimentation in space exploration, pushing the boundaries of scientific knowledge and fueling further interest in the mysteries of the universe. One of the key applications of AI in space exploration is in autonomous spacecraft and rovers. These robotic systems are equipped with AI capabilities that enable them to navigate, analyze, and make decisions without constant human intervention. By leveraging AI, these autonomous machines can adapt to unforeseen circumstances, explore challenging terrain, and execute complex tasks with precision. This has significantly enhanced the efficiency and success of space missions, allowing for more ambitious and long-term exploration endeavors. The ability of autonomous robots to operate independently in extreme environments such as Mars or deep space opens up new possibilities for scientific research and discovery. As the capabilities of AI continue to advance, the future of space exploration holds even greater promise. AI-powered systems have the potential to revolutionize how we explore and colonize other planets, mine asteroids for resources, and even search for extraterrestrial life. The integration of AI in space missions will drive innovation in robotics, machine

147

learning, and data analysis, pushing the boundaries of what is possible in our quest to understand the cosmos. As we venture further into the unknown reaches of space, ethical considerations surrounding the use of AI in space exploration must be carefully addressed to ensure that these technological advancements benefit humanity as a whole.

Robotics and AI in Space Missions

Advances in artificial intelligence have significantly impacted space missions, with the integration of robotics and AI technologies revolutionizing the way we explore the cosmos. Autonomous robots have become essential in performing tasks that are either too dangerous or impractical for human astronauts. These robots can navigate harsh environments, collect data, and execute complex operations with precision. By using AI algorithms, these robots are capable of adapting to unforeseen situations, making split-second decisions, and working efficiently without human intervention. This level of autonomy increases the efficiency and success rate of space missions, ultimately expanding our understanding of the universe. Robotics and AI have enabled the development of innovative spacecraft that can autonomously navigate through space, perform maintenance tasks, and even repair themselves. These advancements have reduced the reliance on ground control and increased the autonomy of spacecraft, allowing them to complete missions more independently. In addition to enhancing the capabilities of spacecraft, robotics and AI also support the exploration of planetary surfaces. Rovers equipped with AI systems can analyze terrain, detect hazards, and collect samples without constant human oversight. This level of sophistication has paved the way for more extensive and in-depth explorations of distant planets and moons. The integration of AI in space missions has opened up new opportunities for scientific research and discovery. By using machine learning algorithms, scientists can analyze vast amounts of data collected from space probes, telescopes, and other instruments to uncover new insights about the universe. AI also plays a crucial role in predicting space weather, identifying potential threats to spacecraft, and ensuring the safety of astronauts during long-duration missions. As we continue to push the boundaries of space exploration, the synergy between robotics and AI will be instrumental in advancing our knowledge of the cosmos and unlocking the mysteries of the universe.

Autonomous Systems for Space Exploration

One of the most significant advancements in the field of artificial intelligence is the development of autonomous systems for space exploration. These systems have the potential to revolutionize the way we explore and understand the universe beyond our planet. Autonomous systems are equipped with the ability to make decisions and carry out tasks without direct human intervention, which is crucial for navigating the complexities of space missions. By utilizing AI technology, these systems can analyze vast amounts of data, adapt to changing conditions, and respond to unforeseen challenges in real-time, making them invaluable for space exploration endeavors. Autonomous systems for space exploration have already demonstrated their capabilities in various missions, such as the Mars rovers and spacecraft missions to distant planets. These systems are designed to operate in extreme environments where human presence is not feasible, making them essential for gathering valuable scientific data and conducting experiments in outer space. The integration of AI algorithms and machine learning techniques allows these systems to learn from their experiences, improve their performance over time, and optimize their decision-making process to achieve mission objectives efficiently and effectively. The development and deployment of autonomous systems for space exploration present unique opportunities and challenges for the future of space exploration. While these systems have the potential to significantly enhance our understanding of the cosmos and advance scientific research, they also raise ethical and safety concerns. Ensuring the reliability, security, and ethical use of autonomous systems in space missions will be crucial to maximizing their benefits and minimizing potential risks. As we continue to push the boundaries of space exploration, autonomous systems powered by AI will play a pivotal role in shaping the future of humanity's exploration of the cosmos.

Ethical Considerations in AI for Space Exploration

Ethical considerations in the development and use of artificial intelligence for space exploration play a critical role in shaping the future of this rapidly advancing technology. As we push the boundaries of AI to navigate extraterrestrial environments and conduct complex missions in space, ethical guidelines must be established to ensure that these systems prioritize safety, fairness, and accountability. Issues such as data privacy, transparency in decision-making, and the potential for bias in AI algorithms must be carefully addressed to uphold ethical standards in the exploration of space. In the context of space exploration, ethical considerations become even more paramount as the consequences of AI errors or malfunctions can have far-reaching impacts on missions, crew members, and the overall success of space endeavors. The stakes are high when relying on AI systems for critical tasks such as navigation, communication, and resource management in the harsh environment of space. Ensuring the robustness and reliability of AI algorithms becomes not only a technical challenge but also an ethical imperative to safeguard the lives and resources involved in space missions. Ethical discussions surrounding AI in space exploration extend beyond technical considerations to encompass broader societal implications. As AI technologies become increasingly integrated into space missions, questions of governance, accountability, and the distribution of benefits and risks arise. Addressing these ethical dilemmas requires a multidisciplinary approach that considers the perspectives of diverse stakeholders, including scientists, engineers, policymakers, and the public. By engaging in open dialogue and collaborative decision-making processes, we can collectively shape a future in which AI for space exploration is developed and utilized in a responsible and ethically sound manner.

XXVIII. AI IN MENTAL HEALTH

As artificial intelligence continues to advance, one of the most promising applications is in the field of mental health. AI has shown great potential in revolutionizing the way mental health services are delivered, offering personalized and accessible support to individuals in need. By utilizing AI-powered chatbots and virtual therapists, individuals can receive immediate guidance and interventions, bridging the gap between demand for mental health services and the availability of professionals. This technology can provide continuous support and monitoring, offering insights into patterns of behavior and emotions that can inform personalized treatment plans. AI in mental health has the potential to reduce stigma associated with seeking help for mental health issues. By offering anonymous and non-judgmental support through virtual platforms, individuals may feel more comfortable opening up about their struggles and seeking assistance. This can lead to early intervention and prevention of more severe mental health issues. AI can help therapists and mental health professionals by providing data-driven insights and recommendations based on patterns identified through analysis of vast amounts of data. This can enhance the effectiveness of treatment plans and improve outcomes for individuals seeking mental health support. Despite the numerous benefits of AI in mental health, there are also ethical considerations that must be taken into account. Privacy concerns, data security, and the potential for bias in algorithms are all important factors that need to be addressed in the development and implementation of AI-powered mental health services. Stakeholders must work together to ensure that ethical guidelines are in place to protect the rights and well-being of individuals accessing these services. By navigating these challenges thoughtfully, the integration of AI in mental health has the potential to significantly improve the quality and accessibility of mental health care for individuals worldwide.

AI-based Mental Health Diagnosis

As artificial intelligence continues to advance, one area that has shown significant potential is AI-based mental health diagnosis. Using machine learning algorithms and natural language processing, AI can analyze patterns in speech, text, and behavior to detect signs of mental health disorders. This innovative approach offers the promise of early intervention and personalized treatment plans for individuals struggling with mental health issues. By providing more accurate and timely diagnoses, AI-based mental health tools have the potential to improve outcomes and reduce the stigma associated with seeking help. AI-based mental health diagnosis tools can augment the capabilities of healthcare providers by offering support in screening and assessment processes. These tools can analyze vast amounts of data to identify subtle changes in behavior or speech that may indicate underlying mental health conditions. By automating routine tasks and providing insights based on data-driven analysis, AI can help clinicians make more informed decisions and tailor treatment plans to the specific needs of each patient. This synergy between human expertise and AI technology has the potential to revolutionize the field of mental health care. The integration of AI-based mental health diagnosis tools also raises important ethical considerations. Privacy concerns, data security, and the potential for bias in AI algorithms are critical issues that must be carefully addressed. Transparency in the development and use of AI tools, as well as robust regulations to safeguard patient confidentiality and data protection, are essential to ensure the ethical use of AI in mental health care. By navigating these challenges thoughtfully, we can harness the power of AI to enhance mental health diagnosis and treatment while upholding the values of integrity, privacy, and equity in healthcare.

Virtual Therapists and AI Counseling

The integration of virtual therapists and AI counseling represents a significant advancement in the field of mental health care. These virtual therapists, powered by artificial intelligence, have the capacity to provide support and guidance to individuals in need of mental health services. By utilizing sophisticated algorithms and natural language processing, virtual therapists can engage in conversations with users, offering therapeutic interventions and coping strategies. This access to round-the-clock mental health support can be particularly beneficial for individuals who may not have easy access to traditional therapy or struggle with stigma associated with seeking help. Virtual therapists have the potential to cater to the unique needs and preferences of each individual, providing personalized interventions that are tailored to their specific mental health concerns. Through continuous interaction and data analysis, these AI-driven platforms can adapt their responses and recommendations based on the user's feedback and progress. This level of customization can enhance the effectiveness of therapy interventions, ultimately improving mental health outcomes for users. The anonymity and privacy afforded by virtual therapists can also help individuals feel more comfortable and open up about their emotions and struggles, leading to more meaningful therapeutic interactions. While virtual therapists and AI counseling hold great promise in expanding access to mental health support, it is essential to address concerns regarding the ethical implications and limitations of these technologies. Issues such as data privacy, confidentiality, and the potential for biases in AI algorithms must be carefully considered to ensure the delivery of ethical and high-quality mental health care. As the field of virtual therapy continues to evolve, ongoing research and regulation will be crucial in safeguarding the well-being of users and maintaining the integrity of mental health services in the digital age.

Privacy and Confidentiality in AI-driven Mental Health Services

As artificial intelligence continues to revolutionize the field of mental health services, the issues of privacy and confidentiality have become paramount. With AI-driven platforms being increasingly utilized for therapy, counseling, and diagnostic purposes, the sensitive data shared by individuals must be safeguarded to maintain trust and ensure ethical practices. Patients often divulge personal information, feelings, and experiences during sessions with AI-driven mental health services, raising concerns about data protection and privacy breaches. Ensuring privacy and confidentiality in AI-driven mental health services requires adherence to stringent security measures and ethical standards. Encryption of data, secure storage protocols, and limited access to confidential information are essential components of safeguarding patient privacy. Clear policies on data sharing and consent must be in place to protect individuals from unauthorized use of their personal information. The development and implementation of robust privacy policies can help build trust between users and AI platforms, ultimately promoting the adoption of these services for mental health support. As AI-driven mental health services continue to evolve, it is imperative for providers and developers to prioritize privacy and confiden-tiality. By establishing strong security protocols, ethical guide-lines, and transparent data practices, the potential benefits of AI in mental health can be maximized while minimizing risks to individual privacy. Adhering to strict privacy standards not only ensures compliance with regulations but also upholds the ethical responsibility of protecting the sensitive information shared by individuals seeking mental health support through AI-driven platforms.

XXIX. AI IN DISASTER RESPONSE

Advancements in artificial intelligence technology have greatly improved disaster response systems, enabling faster and more efficient actions during crises. By utilizing AI algorithms, drones equipped with sensors can quickly assess the impact of disasters, such as earthquakes or floods, and provide real-time data to aid in decision-making. This rapid assessment allows emergency responders to allocate resources more effectively and prioritize areas in need of immediate attention. AI-powered predictive analytics can forecast disaster scenarios based on historical data, weather patterns, and other relevant information, enabling proactive measures to be taken to mitigate risks and potential damage. AI can enhance communication and coordination among various agencies involved in disaster response efforts. Through automated translation services and natural language processing tools, AI can bridge language barriers and facilitate seamless communication between international relief teams and local responders. This ensures a more coordinated and efficient deployment of resources and personnel to affected areas. AI-powered chatbots can provide real-time updates and guidance to those impacted by disasters, offering crucial information on evacuation routes, emergency shelters, and medical facilities. The integration of AI in disaster response systems has the potential to revolutionize the way emergencies are managed and mitigated. By leveraging AI technologies such as machine learning, predictive analytics, and natural language processing, responders can make more informed decisions, prioritize resources effectively, and communicate efficiently during crises.

As AI continues to evolve, it is essential for policymakers, emergency management agencies, and technology developers to collaborate and ensure that these innovative tools are ethically deployed, regulated, and accessible to maximize their impact in saving lives and reducing the overall impact of disasters.

AI for Early Warning Systems

The integration of artificial intelligence in early warning systems has revolutionized the way we approach disaster preparedness and response. By leveraging AI algorithms and machine learning techniques, early warning systems can analyze large volumes of data in real-time to detect patterns and anomalies that may indicate an impending crisis. These systems have the potential to provide timely alerts and predictive insights, allowing authorities to take proactive measures to mitigate risks and protect vulnerable populations. AI-powered early warning systems are instrumental in enhancing disaster resilience and response capabilities, ultimately saving lives and reducing the impact of catastrophic events. Incorporating AI into early warning systems also enables the automation of complex decision-making processes, optimizing resource allocation and response strategies. By utilizing predictive analytics and scenario modeling, AI algorithms can simulate various scenarios and recommend the most effective courses of action based on available data. This predictive capability empowers emergency responders and policymakers to make informed decisions quickly and efficiently, improving overall response coordination and effectiveness. The ability of AI to process and analyze vast amounts of data in real-time enhances situational awareness and enables a more proactive and adaptive approach to disaster management. Despite the numerous benefits of AI for early warning systems, there are also challenges and ethical considerations that must be addressed. Ensuring the reliability and accuracy of AI algorithms, addressing data privacy and security concerns, and mitigating the potential for algorithmic bias are critical aspects that require careful attention. The ethical implications of automated decision-making in high-stakes situations raise questions about accountability, transparency, and human oversight. As AI continues to evolve and integrate into critical infrastructure such as early warning systems, it is imperative to establish robust governance frameworks and ethical guidelines to govern the responsible deployment and use of AI technologies in disaster management.

Robotics in Search and Rescue Operations

One of the most promising applications of artificial intelligence technology today is its use in search and rescue operations through the deployment of robotics. Robotics in search and rescue operations serve as a vital tool in scenarios where human intervention may be limited or too dangerous. These robots are designed to navigate complex terrains, detect survivors, and provide critical information to rescuers. By effectively leveraging robotics, search and rescue teams can enhance their capabilities and improve the likelihood of successful outcomes in challenging situations. The integration of robotics in search and rescue operations has significantly improved the efficiency and effectiveness of emergency response efforts. These robots can be equipped with advanced sensors, cameras, and communication devices to gather real-time data and relay important information to rescue teams. By utilizing robotic technology, search and rescue operations can be conducted more swiftly and accurately, ultimately increasing the chances of locating and rescuing individuals in distress. The use of robots in these operations can also help minimize the risks faced by human responders in hazardous environments, thus ensuring the safety of both rescuers and survivors. While the potential benefits of utilizing robotics in search and rescue operations are substantial, there are also challenges that must be addressed to maximize the effectiveness of these technologies. Issues such as the integration of AI algorithms, the autonomy of robots in decision-making processes, and the interoperability with existing rescue systems need to be carefully considered. Ethical concerns, such as privacy issues and the potential impact on human labor, must be taken into account. By recognizing and addressing these challenges, the full potential of robotics in search and rescue operations can be realized, leading to more efficient, safer, and ultimately more successful outcomes in emergency situations.

Ethical Considerations in AI Deployment during Disasters

As the deployment of artificial intelligence becomes increasingly prevalent, particularly in disaster response scenarios, ethical considerations take on a heightened significance. One primary concern revolves around the potential biases embedded in AI algorithms, which could inadvertently exacerbate existing disparities during crises. If an AI system is programmed with biased data, it may inadvertently prioritize resources towards certain demographics over others, leading to further inequities in access to critical aid during disasters. Issues of transparency and accountability become paramount when AI systems are entrusted with decision-making in high-stakes situations. The lack of transparency in how AI arrives at decisions can undermine trust in the technology and create challenges in ensuring accountability for outcomes. The widespread deployment of AI in disaster response raises questions about the level of human oversight and control necessary to mitigate risks and ensure ethical conduct. While AI systems can process vast amounts of data and make rapid decisions, there is a need for human intervention to validate the outputs and ensure that decisions align with ethical principles and values. As such, striking a balance between the autonomy of AI systems and the involvement of human judgment becomes essential in maintaining ethical standards during disasters. The potential for AI to perpetuate or amplify biases must be addressed through robust oversight mechanisms and continuous monitoring to prevent unintended consequences. Navigating the ethical considerations surrounding AI deployment in disaster scenarios requires a comprehensive approach that encompasses transparency, accountability, and human oversight. By actively addressing biases in AI algorithms, ensuring transparency in decision-making processes, and establishing clear lines of accountability, stakeholders can work towards harnessing the full potential of AI while upholding ethical standards. The responsible deployment of AI in disaster response necessitates a collaborative effort among policymakers, technologists, and ethical experts to develop frameworks that prioritize the well-being and equity of all individuals affected by disasters.

161

XXX. AI IN SPORTS

As artificial intelligence continues to advance, its integration into the world of sports has opened up new opportunities and challenges. One area where AI is making a significant impact is performance analysis. Through the use of AI-powered tools, coaches and athletes can gather and analyze vast amounts of data to improve training techniques, optimize performance, and gain a competitive edge. Sensors embedded in equipment can track an athlete's movements in real-time, providing valuable insights into their technique and physical condition. This data can then be analyzed using AI algorithms to identify patterns, trends, and areas for improvement. AI is revolutionizing the fan experience in sports. With the development of AI-driven technologies such as virtual reality (VR) and augmented reality (AR), fans can now enjoy immersive and interactive experiences from the comfort of their homes. VR technology allows fans to watch games from different camera angles, participate in virtual simulations, and even interact with players in a virtual environment. This not only enhances the entertainment value of sports but also provides new revenue streams for teams and leagues through virtual ticket sales and sponsorships. In addition to performance analysis and fan engagement, AI is also playing a crucial role in injury prevention and player safety in sports. By analyzing biomechanical data and monitoring player movements, AI algorithms can help identify potential injury risks, develop personalized training programs, and even predict when an athlete may be at risk of injury. This proactive approach to player health not only prolongs careers but also ensures the well-being of athletes both on and off the field. As AI continues to evolve,

162

its impact on sports is likely to grow, shaping the future of the industry in ways we have yet to imagine.

Performance Analysis and Prediction

As technology continues to advance, the ability to analyze and predict performance in artificial intelligence systems becomes increasingly important. Performance analysis involves monitoring and evaluating the effectiveness and efficiency of AI algorithms and systems in achieving their intended goals. By examining factors such as processing speed, accuracy, and scalability, researchers can identify areas for improvement and optimization. Performance prediction plays a crucial role in anticipating future outcomes and trends based on historical data and model simulations. One key aspect of performance analysis in AI is the evaluation of algorithm performance in relation to specific tasks or applications. By benchmarking algorithms against established metrics and standards, researchers can assess their effectiveness in solving real-world problems. Performance analysis can help identify bottlenecks or areas of inefficiency within an AI system, leading to targeted improvements and optimizations. This iterative process of analysis and refinement is essential for ensuring that AI technologies continue to evolve and meet the changing demands of users and industries. In the realm of performance prediction, machine learning techniques such as predictive modeling and time series analysis play a critical role in forecasting future outcomes and trends. By leveraging historical data and patterns, researchers can develop predictive models that help anticipate potential issues, identify emerging opportunities, and optimize decision-making processes. Performance prediction can enable proactive maintenance and re-source allocation strategies, leading to improved system relia-bility and operational efficiency. The integration of performance analysis and prediction in AI systems is essential for driving in-novation and achieving optimal outcomes in various domains.

Athlete Training and Optimization

The optimization of athlete training is a crucial aspect of max-imizing performance and achieving competitive success in sports. With the advancement of technology and the integration of artificial intelligence in athlete training programs, coaches and trainers have access to powerful tools that can help in an-alyzing data, identifying patterns, and customizing training re-gimes to suit individual athletes' needs. AI can process vast amounts of data from wearable devices, sensors, and video analysis to provide insights into an athlete's performance, re-covery, and injury prevention. By fine-tuning training programs based on AI recommendations, coaches can help athletes reach their peak physical condition and improve their overall perfor-mance on the field. AI-driven training programs can also help in optimizing training schedules, managing fatigue levels, and pre-dicting potential injuries. By analyzing data on an athlete's sleep patterns, heart rate variability, and nutrition intake, AI algo-rithms can provide recommendations on rest days, recovery strategies, and personalized nutrition plans. This personalized approach to athlete training can lead to better performance out-comes, reduced risk of injuries, and improved overall well-being. AI can simulate game scenarios, create virtual training environ-ments, and provide real-time feedback during practice sessions, enabling athletes to refine their skills and decision-making abilities under pressure. The integration of AI in athlete training programs can also lead to advancements in sports science research, innovation in equipment design, and the development of new training methodologies. By leveraging AI technologies in sports analytics, biomechanics, and performance monitoring, coaches and trainers can stay ahead of the competition and continuously improve their training methods. The evolution of AI in athlete training represents a paradigm shift in how sports are approached, with data-driven insights and personalized feedback playing a pivotal role in optimizing athletic performance and pushing the boundaries of human potential in sports.

Ethical Considerations in AI-enhanced Sports

As artificial intelligence continues to advance, its integration into sports has raised important ethical considerations. One key issue in the realm of AI-enhanced sports is the potential for unfair advantages. As teams or athletes use AI algorithms to analyze opponents, optimize training routines, or even make in-game decisions, there is a risk that those with greater access to advanced AI technology may gain a competitive edge. This could lead to a disparity in performance levels, ultimately undermining the principles of fair play and sportsmanship. It is crucial to establish guidelines and regulations to ensure a level playing field in AI-enhanced sports. Another ethical concern in the context of AI-enhanced sports is the invasion of privacy. With the collection of vast amounts of data on athletes' performance, health metrics, and personal information, there is a risk of this data being misused or exploited. Athletes may feel pressured to share sensitive data in order to stay competitive, raising questions about consent and data security. The use of AI in sports broadcasting to track and analyze player movements raises concerns about surveillance and the boundaries between professional performance and personal privacy. It is essential for stakeholders in the sports industry to address these issues proactively and prioritize the protection of athletes' rights and privacy. The ethical implications of AI-enhanced sports extend to societal values and norms. As AI algorithms increasingly shape how sports are played, coached, and consumed, there is a need to consider the impact on traditional practices and cultural values. The use of AI in refereeing decisions may challenge the human element of sports officiating, raising questions about the role of subjectivity and judgement in sportsmanship. The automation of scouting and talent identification processes through AI may change the dynamics of talent development and recruitment in sports. It is essential to engage in critical dialogue and discourse on the ethical implications of AI in sports to ensure that technology is used in ways that align with societal values and promote the integrity of sports competition.

XXXI. AI IN ENTERTAINMENT INDUSTRY

Advancements in artificial intelligence have significantly impacted the entertainment industry, revolutionizing the way content is created, distributed, and consumed. From personalized recommendations on streaming platforms to the use of AI-powered algorithms for content curation and editing, the integration of AI has optimized the entertainment experience for users worldwide. One notable application of AI in the industry is the development of virtual influencers, digitally created personas that engage with audiences on social media and promote brands. These virtual influencers are designed to resonate with younger demographics and have gained popularity as a marketing strategy in the digital age. AI has also enabled the creation of hyper-realistic computer-generated characters and environments in movies, television shows, and video games. The use of AI-driven visual effects has raised the bar for visual storytelling, blurring the line between reality and fiction. AI tools such as facial recognition software and sentiment analysis algorithms have been utilized by entertainment companies to gauge audience reactions and preferences, helping them tailor their content for maximum engagement. By harnessing the power of AI, the entertainment industry has been able to explore new creative possibilities and enhance the overall entertainment experience for consumers. Looking ahead, the continued advancements in AI technology are poised to shape the future of the entertainment industry even further. As AI algorithms become more sophisticated and capable of understanding and predicting human behavior, content creators will have access to powerful tools for storytelling and audience engagement. The rise of autonomous robots in entertainment, such as AI-powered animatronics and interactive exhibits, will offer immersive experiences for audiences in theme parks, museums, and other entertainment venues. The integration of AI in the entertainment industry represents a paradigm shift in how content is produced, consumed, and experienced, paving the way for a new era of creativity and innovation.

Content Creation and Personalization

The evolution of artificial intelligence has paved the way for significant advancements in content creation and personalization. Virtual assistants, like Siri, Alexa, and Google Assistant, have been instrumental in this evolution by offering personalized services based on users' preferences and behavior. These virtual assistants have the ability to understand and respond to natural language commands, improving their effectiveness in providing tailored recommendations and assistance. As a result, they have become indispensable tools in various industries, from customer service to healthcare, enhancing user experience and driving productivity. With advancements in natural language processing and the integration of machine learning and neural networks, virtual assistants have become increasingly sophisticated in their ability to personalize content. They can analyze vast amounts of data to predict user preferences and behaviors, delivering personalized recommendations and responses in real-time. This level of personalization not only improves user satisfaction but also helps businesses better understand their customers and target their marketing efforts more effectively. As virtual assistants continue to evolve, the line between human-generated and AI-generated content will blur, leading to new possibilities in content creation and personalization. The transition from virtual assistants to autonomous robots represents the next frontier in artificial intelligence, where personalized content creation and services will be taken to new heights. Autonomous robots, such as autonomous vehicles and cleaning robots, are capable of performing tasks independently without human intervention. These robots can adapt to changing environments, learn from their interactions, and provide personalized services tailored to individual needs. The development of autonomous robots also poses technological and ethical challenges, such as concerns about job displacement and safety risks. Despite these challenges, the potential benefits of autonomous robots in revolutionizing industries like medicine, logistics, and agriculture are vast, highlighting the transformative impact of AI on content creation and personalization.

Virtual Reality and AI Integration

Virtual reality and artificial intelligence integration represents a significant advancement in the field of technology. By combining the immersive experience of virtual reality with the intelligent capabilities of AI, new possibilities are unlocked for various industries. One key benefit of this integration is the enhanced training and simulation opportunities it offers. In the medical field, surgeons can practice complex procedures in a virtual environment, allowing them to hone their skills without the need for physical patients. This not only improves patient safety but also reduces costs associated with traditional training methods. The integration of virtual reality and AI can revolutionize the way we interact with technology. Virtual assistants, already familiar to users in the form of Alexa or Google Assistant, could be enhanced with virtual reality capabilities to provide even more personalized and immersive experiences. Imagine being able to have a virtual assistant guide you through a virtual tour of a new city or help you practice a presentation in a simulated environment. The possibilities for enhancing user engagement and experience are truly limitless with this integration. The use of virtual reality and AI integration in industries such as gaming and entertainment can create entirely new experiences for users. The ability to interact with intelligent virtual characters in a realistic and immersive environment opens up new avenues for storytelling and gameplay. From virtual reality escape rooms that adapt to the player's decisions to AI-controlled characters that respond to the player's emotions, the integration of these technologies can elevate entertainment experiences to new heights. As technology continues to evolve, the potential for virtual reality and AI integration to shape the future of various industries is both exciting and promising.

Impact on Audience Engagement and Experience

The impact of virtual assistants on audience engagement and experience cannot be understated. These AI-powered tools have revolutionized the way individuals interact with technology on a daily basis. By providing users with a seamless and intuitive interface for tasks like setting reminders, checking the weather, or ordering food, virtual assistants have become an indispensable part of modern life. The convenience and efficiency they offer have significantly enhanced user engagement and satisfaction across various industries, leading to a more personalized and interactive experience for consumers. As technology continues to evolve, the transition from virtual assistants to autonomous robots represents a significant advancement in the field of artificial intelligence. Autonomous robots, with their ability to perform tasks independently and make decisions based on environmental cues, have the potential to further enhance audience engagement and experience. From autonomous vehicles navigating city streets to cleaning robots tidying up homes, these machines are reshaping the way we interact with our surroundings. The immersive and interactive nature of autonomous robots has the power to captivate audiences and create a truly unique and engaging experience that blurs the lines between human and machine interaction. The widespread adoption of autonomous robots also raises important ethical and societal considerations. As these machines become more integrated into everyday life, concerns about privacy, safety, and job displacement need to be addressed. By carefully navigating these challenges and implementing appropriate regulations, we can ensure that the impact of autonomous robots on audience engagement and experience remains positive and beneficial. The evolution of artificial intelligence from virtual assistants to autonomous robots has the potential to transform the way we live, work, and interact with technology, opening up new possibilities for innovation and growth in the digital age.

XXXII. AI IN CUSTOMER SERVICE

The integration of artificial intelligence in customer service has revolutionized the way businesses interact with their clients. Through the use of AI-powered chatbots and virtual assistants, companies can provide immediate and personalized assistance to customers, enhancing their overall experience. These AI tools are not only capable of handling basic inquiries and tasks but can also analyze customer data to offer tailored recommendations and solutions. By leveraging AI in customer service, organizations can improve efficiency, reduce response times, and increase customer satisfaction levels. One of the key advantages of AI in customer service is its ability to handle a high volume of incoming queries simultaneously, without the need for human intervention. This not only reduces the burden on customer service representatives but also ensures that customers receive prompt and accurate responses to their inquiries. AI-powered systems can continuously learn and improve their performance over time, adapting to customer preferences and evolving trends. This adaptive nature of AI allows businesses to stay ahead of the curve and provide a seamless and consistent customer experience across various touchpoints. AI in customer service enables companies to gather valuable insights from customer interactions, which can be used to enhance products, services, and marketing strategies. By analyzing patterns in customer behavior and sentiment, organizations can identify areas for improvement, anticipate customer needs, and personalize their offerings. This proactive approach not only strengthens customer relationships but also drives business growth and competitiveness in today's digital landscape. As AI technology continues to evolve, the role of AI in customer service will undoubtedly expand, offering endless possibilities for innovation and better customer engagement.

Chatbots and Automated Support Systems

As technology continues to advance, chatbots and automated support systems have become increasingly prevalent in various industries. These AI-driven tools are designed to provide assistance and information to users, streamlining processes and enhancing customer service. Virtual assistants, such as Siri and Alexa, were among the first iterations of chatbots, paving the way for more sophisticated automated support systems. These virtual assistants revolutionized the way individuals interact with technology by allowing for voice commands and natural language processing capabilities, marking a significant milestone in the evolution of AI. With the advancements in natural language understanding and machine learning, chatbots have evolved to offer more personalized and efficient support. The integration of neural networks has improved the accuracy and responsiveness of automated systems, enabling them to handle complex queries and tasks. This progress has not only enhanced user experience but has also increased the adoption of chatbots in sectors ranging from customer service to healthcare. The ability of automated support systems to interpret and respond to human language has had a profound impact on how businesses engage with their customers, leading to greater efficiency and customer satisfaction. Moving forward, the development of autonomous robots represents the next frontier in artificial intelligence. These robots, equipped with advanced sensors and decision-making capabilities, have the potential to revolutionize

173

industries such as manufacturing, transportation, and healthcare. The widespread adoption of autonomous robots also raises ethical and societal concerns, including issues related to job displacement and data privacy. It is essential for policymakers and industry stakeholders to address these challenges proactively to ensure that the benefits of autonomous robots are maximized while mitigating potential risks.

Personalized Customer Interactions

As artificial intelligence continues to revolutionize various industries, personalized customer interactions have become a key focus for businesses seeking to enhance user experiences and drive customer loyalty. Through the implementation of AI-powered virtual assistants, companies can now offer tailored support and recommendations based on individual preferences and behaviors. These virtual assistants, such as Siri, Alexa, and Google Assistant, have evolved to understand and respond to complex commands, making them valuable tools for delivering personalized services to customers in real-time. The advancements in natural language processing have played a crucial role in enhancing the capabilities of virtual assistants, enabling them to engage in more sophisticated conversations and provide more accurate responses. By integrating machine learning and neural network technologies, virtual assistants can now adapt to individual preferences, learn from past interactions, and anticipate user needs. This level of personalization not only enhances the overall user experience but also helps businesses improve customer satisfaction and drive sales by delivering targeted recommendations and offers. Looking ahead, the development of autonomous robots represents the next frontier in artificial intelligence, offering even more opportunities for personalized customer interactions. Autonomous robots, such as autonomous vehicles, cleaning robots, and robots in industry, have the potential to revolutionize customer service by providing seamless and efficient assistance in various settings. As businesses explore the possibilities of autonomous robots, they must also navigate the technological and ethical challenges associated with their development to ensure that these robots can deliver personalized experiences while upholding ethical standards and data privacy regulations.

Challenges and Opportunities in AI-driven Customer Service

When considering the evolution of artificial intelligence, one cannot overlook the significant impact that AI-driven customer service has had on businesses and consumers alike. One of the key challenges in implementing AI-driven customer service is the potential loss of the personal touch that traditional customer service interactions provide. Customers may feel disconnected or frustrated when interacting with a machine rather than a human, leading to decreased satisfaction and loyalty. This challenge also presents an opportunity for companies to enhance their AI systems to provide more personalized and empathetic responses. By incorporating sentiment analysis and emotional recognition technology, AI-driven customer service can improve the quality of interactions and better meet the needs of customers. Another challenge in AI-driven customer service is the risk of algorithmic biases impacting the way individuals are served. AI systems are only as good as the data they are trained on, and if this data is biased or incomplete, it can lead to discriminatory outcomes. Companies must be vigilant in monitoring and mitigating biases in their AI systems to ensure fair and equitable customer service experiences. On the other hand, this challenge also opens up opportunities for companies to leverage AI to proactively address biases and promote diversity and inclusion in their customer service operations. By actively engaging in algorithmic transparency and fairness initiatives, businesses can build trust with their customers and demonstrate a commitment to ethical AI practices. Despite the challenges, AI-driven customer service presents numerous opportunities for businesses to streamline operations, improve efficiency, and enhance customer experiences. By leveraging AI technologies such as natural language processing, machine learning, and predictive analytics, companies can automate routine tasks, provide faster response times, and deliver personalized recommendations to customers. AI-driven customer service allows companies to gather and analyze vast amounts of data to gain valuable insights into customer behavior and preferences. With the right strategies in place, businesses can harness the power of AI to optimize their customer service processes and drive growth and success in an increasingly competitive market.

XXXIII. CONCLUSION

As the journey from virtual assistants to autonomous robots unfolds, it becomes evident that artificial intelligence has made significant strides in reshaping various aspects of our lives. The progression from basic virtual assistants like Siri and Alexa to sophisticated autonomous robots represents a monumental shift in the capabilities of AI technologies. These advancements have not only enhanced our efficiency and productivity in different domains but have also raised important ethical considerations that must be addressed. The transition towards autonomous robots marks a new phase in the evolution of AI, where machines are increasingly capable of making decisions and performing tasks autonomously. The real-world applications of autonomous robots span across diverse sectors such as medicine, logistics, agriculture, and security, showcasing the immense potential of AI to revolutionize traditional practices. While the benefits of autonomous robots are undeniable, there are also inherent risks associated with their widespread adoption. As we navigate this rapidly changing landscape, it is crucial to strike a balance between technological innovation and ethical considerations to ensure that AI advancements are leveraged for the greater good of society. The future of artificial intelligence holds immense promise, but it also requires careful navigation to address the complex challenges that lie ahead. The evolution of artificial intelligence from virtual assistants to autonomous robots represents a profound transformation in how we interact with technology and the potential impact it can have on various industries. As we look towards the future of AI, it is imperative to anticipate the trends and perspectives that will shape the landscape of innovation. By reflecting on the lessons learned from the journey so far, we can better prepare for a future where AI will continue to play a central role in shaping our world. Through thoughtful regulation and ethical considerations, we can harness the power of AI to build a more inclusive and sustainable future for all.

Recap of AI Evolution from Virtual Assistants to Autonomous Robots

Technological progress has propelled the evolution of artificial intelligence from virtual assistants to autonomous robots. The journey began with the introduction of virtual assistants like Siri, Alexa, and Google Assistant, which revolutionized the way we interact with technology. These virtual assistants initially focused on basic tasks such as setting reminders and playing music but quickly advanced to handle more complex commands through improvements in natural language processing. Their impact on everyday life and various industries was significant, paving the way for further developments in AI. As speech recognition technology improved, the focus shifted towards enhancing natural language understanding in virtual assistants. This led to the integration of machine learning and neural network technologies, enabling virtual assistants to interpret and respond to more nuanced commands effectively. The exponential growth in the capabilities of virtual assistants laid the foundation for the next frontier in AI evolution - autonomous robots. These robots, characterized by their ability to operate independently and make decisions based on real-time data, have become integral in sectors such as transportation, healthcare, and manufacturing. The development of autonomous robots presents unique challenges, both technological and ethical. Their applications in various sectors such as medicine, logistics, agriculture, and security are driving business efficiency and productivity to new heights. As AI continues to advance, it is essential to anticipate the potential benefits and risks associated with the widespread adoption of autonomous robots. The evolution from virtual assistants to autonomous robots underscores the transformative power of artificial intelligence in reshaping the future of human-machine interactions and the societal landscape at large.

Reflection on the Transformative Impact of AI Technologies

The transformative impact of AI technologies is evident in the evolution from virtual assistants to autonomous robots. Virtual assistants, such as Siri, Alexa, and Google Assistant, marked the initial step in integrating AI into everyday life. These AI-driven technologies have not only streamlined tasks and increased efficiency but also revolutionized how individuals interact with devices and access information. The development of natural language processing has further enhanced the capabilities of virtual assistants, enabling them to understand and respond to complex commands with greater accuracy. The integration of machine learning and neural network technologies has significantly improved the overall performance and responsiveness of virtual assistants, paving the way for more advanced AI applications. As technology progresses, autonomous robots have emerged as the next frontier of artificial intelligence. These robots possess the ability to operate independently, making decisions and carrying out tasks without human intervention. From autonomous vehicles to cleaning robots and industrial robots, these AI-driven machines are revolutionizing various sectors. The development of autonomous robots also raises technological and ethical challenges. Issues such as safety, privacy, and the impact on the workforce need to be carefully addressed to ensure the responsible deployment and regulation of autonomous robots in society. Despite these challenges, the potential benefits of autonomous robots in improving business efficiency, productivity, and safety cannot be overlooked. Reflecting on the transition from virtual assistants to autonomous robots underscores the profound impact that AI technologies have had on society. As AI continues to evolve and become more integrated into various aspects of human life, it is crucial to consider the ethical implications and ensure that technological advancements are aligned with societal values. Striking a balance between innovation and ethical considerations will be key in harnessing the potential of AI to improve lives while mitigating potential risks. The future of artificial intelligence holds immense promise, but thoughtful reflection and proactive regulation are essential to ensure that AI technologies serve the collective good.

Call to Action for Ethical and Responsible AI Development

As the field of artificial intelligence continues to rapidly advance, there is an urgent call for ethical and responsible development practices to be prioritized. The evolution from virtual assistants to autonomous robots signifies a significant shift in the capabilities and impact of AI technologies. With virtual assistants serving as the initial step towards AI integration in daily life, it is essential to consider the ethical implications of these technologies. The development of AI must be guided by principles that prioritize the well-being and safety of individuals, ensuring that these systems are designed and utilized ethically. Moving beyond virtual assistants, the emergence of autonomous robots brings about a new set of challenges and considerations. These robots have the potential to revolutionize various industries and human activities, but their development must be approached with a keen awareness of the ethical implications. From ensuring safety and security to addressing potential job displacement, the responsible integration of autonomous robots requires careful consideration of the societal impacts of these technologies. As such, a call to action for ethical and responsible AI development is essential to navigate the complexities of this evolving technological landscape. The transition from virtual assistants to autonomous robots underscores the need for a proactive approach to ethical AI development. By establishing clear guidelines and regulations, stakeholders can ensure that AI technologies are designed and deployed in a manner that aligns with ethical standards and societal values. As AI continues to play an increasingly prominent role in various aspects of human life, it is imperative that the development of these technologies remains grounded in ethical considerations. Only through a concerted effort to prioritize responsible AI development can we harness the full potential of these technologies while mitigating potential risks and ensuring a positive impact on society.

BIBLIOGRAPHY

Varsha Jain. 'Artificial Intelligence in Customer Service.' The Next Frontier for Personalized Engagement, Jagdish N. Sheth, Springer Nature, 8/17/2023

Regina Lynn Preciado. 'The Personalization Paradox.' Why Companies Fail (and How To Succeed) at Delivering Personalized Experiences at Scale, Val Swisher, XML Press, 3/8/2021

Mr. Ghiath Shabsigh. 'Powering the Digital Economy: Opportunities and Risks of Artificial Intelligence in Finance.' El Bachir Boukherouaa, International Monetary Fund, 10/22/2021

Hayit Greenspan. 'Deep Learning for Medical Image Analysis.' S. Kevin Zhou, Academic Press, 11/23/2023

Kaveh Memarzadeh. 'Artificial Intelligence in Healthcare.' Adam Bohr, Academic Press, 6/21/2020

Yen-Wei Chen. 'Handbook of Artificial Intelligence in Healthcare.' Vol 2: Practicalities and Prospects, Chee-Peng Lim, Springer Nature, 11/26/2021

Florentin Smarandache. 'The Encyclopedia of Neutrosophic Researchers, 1st volume.' Infinite Study, 11/12/2016

Grégoire Montavon. 'Explainable AI: Interpreting, Explaining and Visualizing Deep Learning.' Wojciech Samek, Springer Nature, 9/10/2019

Francesco Petruccione. 'Machine Learning with Quantum Computers.' Maria Schuld, Springer Nature, 10/17/2021

Indrajit Pan. 'Quantum Machine Learning.' Siddhartha Bhattacharyya, Walter de Gruyter GmbH & Co KG, 6/8/2020

Raymond Laflamme. 'An Introduction to Quantum Computing.' Phillip Kaye, OUP Oxford, 1/1/2007

Abhishek Kumar. 'Quantum Computing and Artificial Intelligence.' Training Machine and Deep Learning Algorithms on Quantum Computers, Pethuru Raj, Walter de Gruyter GmbH & Co KG, 8/21/2023

Charles Morgan. 'Responsible AI.' A Global Policy Framework, International Technology Law Association, 1/1/2019

Malik Ghallab. 'Reflections on Artificial Intelligence for Humanity.' Bertrand Braunschweig, Springer Nature, 2/6/2021

Jaime Wood. 'The Word on College Reading and Writing.' Carol Burnell, Open Oregon Educational Resources, 1/1/2020

Ezekiel J. Emanuel. 'Ethical and Regulatory Aspects of Clinical Research.' Readings and Commentary, Johns Hopkins University Press, 1/1/2003

Health and Medicine Division. 'Communities in Action.' Pathways to Health Equity, National Academies of Sciences, Engineering, and Medicine, National Academies Press, 4/27/2017

Marco Antonio Aceves-Fernandez. 'Artificial Intelligence.' Emerging Trends and Applications, BoD – Books on Demand, 6/27/2018

Chen Qiufan. 'AI 2041.' Ten Visions for Our Future, Kai-Fu Lee, Crown, 3/5/2024

Gupta, Brij B.. 'Security, Privacy, and Forensics Issues in Big Data.' Joshi, Ramesh C., IGI Global, 8/30/2019

Juhi Kulshrestha. 'Ethics in Artificial Intelligence: Bias, Fairness and Beyond.' Animesh Mukherjee, Springer Nature, 12/29/2023

Teresa Scantamburlo. 'Machines We Trust.' Perspectives on Dependable AI, Marcello Pelillo, MIT Press, 8/24/2021

Mariya Ouaissa. 'AI and IoT for Sustainable Development in Emerging Countries.' Challenges and Opportunities, Zakaria Boulouard, Springer Nature, 1/31/2022

Christopher C. Nicholls. 'Financial Institutions.' The Regulatory Framework, LexisNexis, 1/1/2008

Bernd Carsten Stahl. 'Artificial Intelligence for a Better Future.' An Ecosystem Perspective on the Ethics of AI and Emerging Digital Technologies, Springer Nature, 3/17/2021

American Nurses Association. 'Code of Ethics for Nurses with Interpretive Statements.' Nursesbooks.org, 1/1/2001

Muzaffar Munshi. 'The Ethics Of Artificial Intelligence: Balancing Benefits and Risks.' Muzaffar Munshi, 5/13/2023

Francis X. Govers. 'Artificial Intelligence for Robotics.' Build intelligent robots that perform human tasks using AI techniques, Packt Publishing Ltd, 8/30/2018

A. Pugh. 'Robot Vision.' Springer Science & Business Media, 6/29/2013

Miomir Vukobratovic. 'Introduction to Robotics.' Springer Science & Business Media, 12/6/2012

Chenyan Xiong. 'Neural Approaches to Conversational Information Retrieval.' Jianfeng Gao, Springer Nature, 3/16/2023

Paul Le Grand. 'The Essential Book on NLP Neurolinguistic Programming.' Unleash Your Unlimited Power with 17 NLP Techniques, Coaching with NLP, NLP for Business, NLP Technology, Mind Control and Emotional Intelligence, Independently Published, 3/30/2021

Nuno Sergio Marques Antunes. 'The Importance of the Tidal Datum in the Definition of Maritime Limits and Boundaries.' IBRU, 1/1/2000

Dan Jurafsky. 'Speech & Language Processing.' Pearson Education, 9/1/2000

Alma Y. Alanis. 'Neural Networks for Robotics.' An Engineering Perspective, Nancy Arana-Daniel, CRC Press, 9/6/2018

Bernhard Mehlig. 'Machine Learning with Neural Networks.' An Introduction for Scientists and Engineers, Cambridge University Press, 10/28/2021

Kevin Gurney. 'An Introduction to Neural Networks.' CRC Press, 10/8/2018

Joachim Reinhardt. 'Neural Networks.' An Introduction, Berndt Müller, Springer Science & Business Media, 10/2/1995

Ilias G. Maglogiannis. 'Emerging Artificial Intelligence Applications in Computer Engineering.' Real Word AI Systems with Applications in EHealth, HCI, Information Retrieval and Pervasive Technologies, IOS Press, 1/1/2007

Sahil Puri. 'Machine Learning and Data Science Blueprints for Finance.' Hariom Tatsat, "O'Reilly Media, Inc.", 10/1/2020

Christoph Molnar. 'Interpretable Machine Learning.' Lulu.com, 1/1/2020

Yoshua Bengio. 'Deep Learning.' Ian Goodfellow, MIT Press, 11/10/2016

Health and Medicine Division. 'Making Eye Health a Population Health Imperative.' Vision for Tomorrow, National Academies of Sciences, Engineering, and Medicine, National Academies Press, 1/15/2017

Vicente Julian. 'Personal Assistants: Emerging Computational Technologies.' Angelo Costa, Springer, 8/29/2017

Rosalie K. Ambler. 'Morale as a Function of Self-definition and Stage of Training.' George M. Rickus, Naval Aerospace Medical Institute, Naval Aerospace Medical Center, 1/1/1967

Nick Loper. 'Virtual Assistant Assistant.' The Ultimate Guide to Finding, Hiring, and Working with Virtual Assistants, CreateSpace Independent Publishing Platform, 8/3/2013

Narayan Changder. 'Research methodology.' The amazing quiz book, Changder Outline, 12/21/2022

Michael Wooldridge. 'A Brief History of Artificial Intelligence.' What It Is, Where We Are, and Where We Are Going, Flatiron Books, 1/19/2021

Christoph Lütge. 'An Introduction to Ethics in Robotics and AI.' Christoph Bartneck, Springer Nature, 8/11/2020

www.ingramcontent.com/pod-product-compliance
Lightning Source LLC
Chambersburg PA
CBHW052316220526
45472CB00001B/140